CO-ANZ-666

The Behaviour of Nematodes

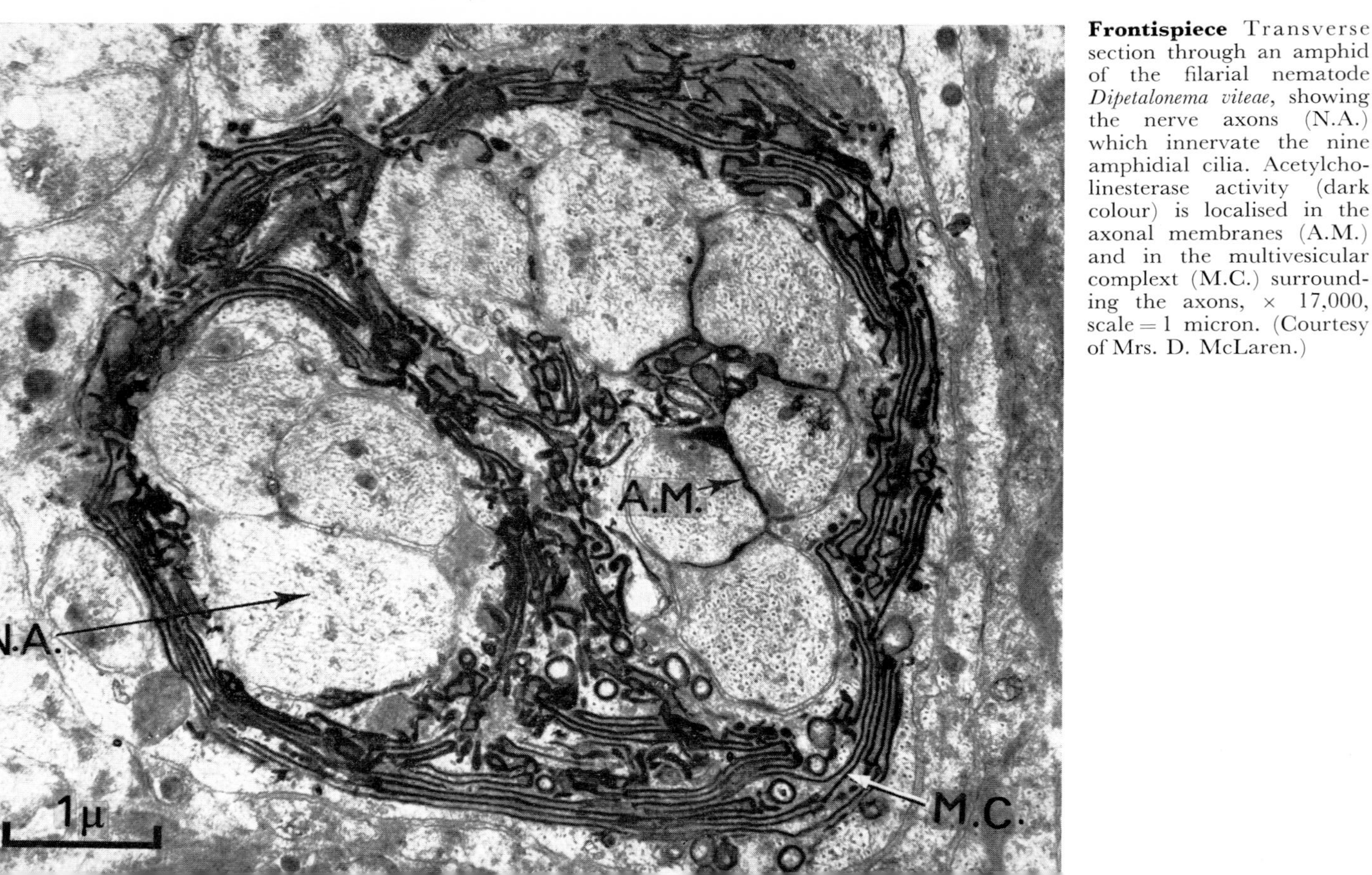

Frontispiece Transverse section through an amphid of the filarial nematode *Dipetalonema viteae*, showing the nerve axons (N.A.) which innervate the nine amphidial cilia. Acetylcholinesterase activity (dark colour) is localised in the axonal membranes (A.M.) and in the multivesicular complext (M.C.) surrounding the axons, × 17,000, scale = 1 micron. (Courtesy of Mrs. D. McLaren.)

The Behaviour of Nematodes

their activity, senses and responses

Neil A. Croll, Ph.D

Department of Zoology, and Applied Entomology,
Imperial College, University of London

Edward Arnold (Publishers) Ltd, London

First published 1970

SBN: 7131 2286 2

Printed in Great Britain by
C. Tinling & Co. Ltd, Prescot, Lancs

Preface

Nematode behaviour has been studied increasingly in recent years and much valuable information is scattered throughout the literature of nematologists and parasitologists. This together with further data usually passed by word of mouth, needs to be brought together in book form. However, my reason for writing this review was not only to collect and consolidate our present knowledge of the subject, but also to indicate its relevance to current research. Another aim was to bring together information on the free living and parasitic nematodes.

In preparing this review, some may think I have moved prematurely into a subject which is still in its infancy. If these pages help, however, to form a background to more sophisticated and exact experimental and field work, I am willing to accept the criticism.

I gratefully acknowledge the following who have granted me permission to reproduce figures and plates: Dr. A. F. Bird, Dr. C. D. Blake, E. J. Brill Ltd., Publishers, Prof. J. J. C. Buckley, Dr. H. D. Crofton, Dr. G. Dean, Dr. C. C. Ellenby, Prof. Sir James Gray, Dr. C. D. Green, Mr. D. N. Greet, Dr. F. Hawking, Mr. J. J. Hesling, Dr. I. K. Ibrahim, Dr. W. G. Inglis, Dr. F. G. W. Jones, Dr. A. R. Maggenti, Miss D. J. McLaren, Dr. D. G. Murphy, Dr. R. S. Pitcher, Mr. J. Robinson, Prof. W. P. Rogers, Dr. H. Streu, Rev. R. W. Timm, Dr. D. R. Viglierchio and Dr. M. Yoeli.

I am especially indebted to Dr. H. R. Wallace, Dr. F. G. W. Jones and Rev. R. W. Timm for many useful comments on the manuscript, and for the permission of the former to use unpublished work. I also thank Dr. H. Streu, for sending me his film on the movement of *Criconemoides curvatum*. Dr. A. F. Bird, Dr. A. R. Maggenti, the late Prof. B. G. Peters, Dr. D. R. Viglierchio, Dr. M. A. Odei, Dr. H. D. Chapman and my wife Doreen, all commented on parts of the script. The assistance of Mrs. Janet Croll and Mr. J. Smith who typed and prepared the figures, respectively, is particularly appreciated.

1970 N. A. CROLL
Imperial College

Contents

1 | APPROACHES TO NEMATODE BEHAVIOUR

Nematodes have occupied a wide range of ecological niches and in addition to parasitizing fungi and higher plants, invertebrates and vertebrates, there are free-living forms in soil, marine and fresh water habitats. There is much morphological uniformity throughout the group and it is differences in behaviour and physiology, superimposed on a sound basic plan which have been instrumental in the success of the Nematoda.

The task of describing and cataloguing the many species of nematodes is incomplete but enough has been done for several alternative systematic schemes to be proposed. While advances have been, and are being made in this direction, new questions are being posed on the ecology, physiology and behaviour of nematodes.

In nematology most of the research has been directed towards the understanding and control of economically important worms. This is reflected in the knowledge of nematode behaviour and most of the available information concerns zoo-parasitic and phyto-parasitic forms.

Many think that the separation of animal-parasitic and plant-parasitic nematodes has impeded the progress of nematology, and that an exchange of ideas would enrich the subject. The study of nematode behaviour would gain by comparing it with that of other phyla, especially some of the parasitic platyhelminths. Digenean larvae, for example, with free-living motile miracidia and cercariae, are simple animals that share the ecological problems of transmission comparable with those of parasitic nematodes.

Nematology is a distinct corner of biology but because of the applied bias in earlier work studies of behaviour have often been incorporated into bionomics and epidemiology. This has led to much scattered information, forming an appendage to an otherwise bionomical or physiological study. The bionomical flavour of the work has supplied much field data but few precise laboratory experiments. Although the study of behaviour is most relevant when related to all the stimuli of natural conditions, isolation of single factors in the laboratory gives more reliable information, leading to an understanding of sensory physiology. Thus, the photosensitivity of *Dirofilaria immitis* adults (page 30), filarial worms living in the heart of dogs, which are never exposed to light, may indicate a basic and widespread photosensitivity among nematodes—an observation that would never have been made in field conditions. Adult *Camallanus* and *Rhabdias bufonis*, parasites of cold-blooded hosts, show a similarly anomalous reaction by responding positively in a thermal gradient (page 48).

The infective stage

The nematode life cycle includes an egg, four larval stages each separated by a moult, and adults which are almost always separate males and females. Examples are known where any or all of the stages are parasitic. In the nematodes most studied, however, the adults are usually parasitic, and the larval stages free-living or parasitic in an intermediate host. One stage has usually been developed which completes the transition from free-living to parasitic—the infective stage. This is the third larval stage or L3 in most Secernentia, but may be the L2 as in *Heterodera* or the L4 as in *Ditylenchus* both phytoparasites. The free-living infective stage in animal parasitic nematodes is often a long-lived, ensheathed, resistant, non-feeding stage, bridging the gap between the free-living and the parasitic habits.

Rogers (1962) summarized the physiological adaptations of the infective stage, the stage which contacts and enters the host. The first and second stage larvae are feeding stages, often remaining in dung and restricted to areas of organic decay. The infective form must be uniquely adapted to live for periods both outside and inside its host. Many of the behavioural adaptations of the infective stage can be interpreted as aids in locating the host. The responses of the different larval stages of parasitic nematodes need to be compared with free-living nematodes, and related to their role in the parasitic life cycle (Croll, 1966d).

Behaviour and systematics

The preservation of favourable mutations by natural selection is accepted in modern biological thought, and fundamental to the theory of the evolutionary mechanism. Such forces have been as significant in the establishment of favourable behavioural responses which are genetically controlled, as in the evolution of structure.

In attempting to organize the information of nematode behaviour into a scheme, it is not surprising to find that a correlation exists between the behaviour and biology of nematodes. The systematics and the biology of nematode life cycles also correspond. In the Secernentia the third stage larva is typically the infective stage; Ancylostomatidae have actively-penetrating larvae; and the Trichostrongylidae and Strongylidae enter their hosts as contaminants of food. Many other correlations between biology and systematics could be cited, but these are used in classification so the correlation is inevitable. It is more reliable to relate behaviour to biology or ecological niche.

As more behavioural responses become known for nematodes, they may be incorporated together with morphological and other data into systematic and phylogenetic thinking but at present this is premature.

Teleology in the description of responses

Herter (1942) showed that the duck leech *Protoclepsis tessalata* was attracted to any object moving through water; this was regarded as an adaptive response which leads the leech to its host. Rothschild (1962) reviewed the effects of parasitism on the behaviour of some infected hosts. She discussed the way in which cotylurid metacercariae encyst in the eyes of fish, making them blind and unable to avoid predatory fish or birds; and the effects of encysted metacercariae in the lateral lines of fish. *Modiolaria* sp., a commensal mollusc of tunicates, seems to be attracted to the protein 'tunicin' (Bourdillon, 1950), and another commensal mollusc *Montacuta ferroginosa* responds chemokinetically to secretions of its host, the heart urchin *Echinocardium cordatum* (Morton, 1962).

There are many similar examples in the literature. Behind these observations is the inference that the parasites' behaviour, or change in the behaviour of the infected host, is advantageous to parasitic transmission. The behaviour of the proglottids of the tapeworm *Taenia solium* and *T. saginata* occurring in human faeces differs in such a way as to enhance the possibility of transmission (Monnig, 1941). *T. saginata* proglottids migrate onto herbage, where they make contact with the cattle hosts, while *T. solium* proglottids remain in the faeces where they are more likely to be eaten by the pig.

If it were stated that metacercariae encysted in the eyes of fish to aid their transmission, the phrase would be teleological and open to criticism. It is foolish to say that migrating proglottids of *T. saginata* are 'looking for' a cow, but it is permissible to say that selection pressures have favoured adaptations that help the proglottid reach the next host.

If the hazards are realized, teleology is a useful short-hand allowing a biological inference to be made, without shrouding the statement in a repetitive preamble about natural selection.

The behaviour of the infective stages of parasitic nematodes, if teleology be permitted, is directed at a single end other than immediate survival, that of reaching the host and entering it.

The responses

Much confusion has occurred in the description of nematode behavioural responses, because of inconsistent adherence to the nomenclature set up for the categories of orientation response. Thus, many authors have reported taxes merely on the basis of a concentration of worms at one or other end of a physical or chemical gradient.

Fraenkel and Gunn (1940) did much to clarify and standardize the terminology in animal behaviour. Ewer and Bursell (1950), Kennedy (1945), and Gunn, Kennedy and Pielou (1937) also contributed to the classification of behaviour patterns in lower organisms and Carthy (1958, 1966) produced standard works on the orientation responses of invertebrates.

For clarity, the two main categories of response and their sub-divisions are listed below; the definitions are largely those of Fraenkel and Gunn (1940), with modifications after Ewer and Bursell (1950).

1 KINESIS

An undirected response to a variation in the stimulus without orientation of the body axis to the source of stimulation; movement resulting from a kinesis is random.

(a) *Orthokinesis*

A non-directional response in which the speed or frequency of activity depends on the intensity of stimulation.

(b) *Klinokinesis*

A non-directional response in which the rate of turning depends on the intensity of stimulation.

Both forms of kinesis can be further divided into those in which sensory adaptation occurs and those without sensory adaptation, each resulting in a different type of orientation (Ewer & Bursell, 1950).

2 TAXIS

A directed orientation, where the long axis of the body is orientated in line with the source of stimulation and locomotion is towards (positive) or away from (negative) the source. A taxis is not a tropism, a term used to describe growth movements of plants and the bending movements of sessile animals such as hydroid coelenterates.

(a) *Klinotaxis*

A tactic response where orientation is by regular lateral deviations of part or the whole of the body, by comparing the intensities of stimulation at successive time intervals. The minimum sensory equipment is a single receptor. Klinotaxes can occur without lateral deviation although this is not likely in nematodes.

(b) *Tropotaxis*

A tactic response where orientation results from turning to the side which is more or less stimulated, by a simultaneous comparison of intensities on either side. The minimum sensory equipment is paired receptors. Ewer and Bursell (1950) subdivide tropotactic orientation into that from bilaterally placed receptors, and that from antero-posteriorly placed receptors. They demonstrated orientation in the onychophore *Peripatopsis moseleyi* where stimulation is compared simultaneously at either end of the body (Bursell & Ewer, 1950) and suggested that this type of orientation may be common in vermiform animals (Ewer & Bursell, 1950).

2 NEMATODE MOVEMENT

With very few exceptions, nematodes progress by sinusoidal or undulatory propulsion (Gray 1953, Wallace 1963, Gray & Lissmann 1964). The nematode body is thrown into dorso-ventral waves, usually initiated at the anterior end, which pass backwards down the body but occasionally starting at the tail and moving forwards (page 96). The mechanism of undulatory movement in nematodes may be studied through the hydrostatic skeleton, the cuticle, and the musculature.

Hydrostatic skeleton

A high internal pressure or hydrostatic skeleton can be demonstrated if a living nematode is pierced or strongly heated. Worms frequently burst and the contents are forced out under pressure. Similarly, when at rest or in a narcotized state many nematodes straighten out although many do coil. These postures are probably associated with the high internal pressure of the hydrostatic skeleton. This pressure was accurately measured in *Ascaris lumbricoides*, with a Bourbon gauge made from a thin glass helix filled with liquid, and an indentation gauge. The pseudocoelomic pressure of an adult female was about 100 mm mercury or 135 cm water. (Atmospheric pressure=760 mm mercury). The mean hydrostatic turgor pressure was 70 mm mercury (95 cm water), but varied widely and often rhythmically from 16 mm to 225 mm (Harris & Crofton 1957).

A hydrostatic skeleton is not unique to the Nematoda, and its distribution and significance in invertebrates was reviewed by Chapman (1958).

The nematode oesophagus, or pharynx, is a muscular cylinder, often with a non-return valve, that pumps food into the intestine against the hydrostatic pressure of the body. Structural adaptations are also found at the anus and vulva, to compensate for the hydrostatic skeleton.

The cuticle

The maintenance of a high internal pressure can only be achieved if the nematode cuticle is strong and sufficiently inelastic to resist expansion.

The cuticle of *Ascaris lumbricoides*, and probably many other nematodes, has a lattice structure, composed of two sets of superimposed fibres, running at such an angle to each other that they enclose a system of minute parallelograms as seen in surface view. These fibres strengthen the cuticle and assist in withstanding the internal pressure and yet allow locomotion.

The diagonals of the parallelograms lie longitudinally and transversely in relation to the axis of the nematode body, so that when the longitudinal muscles contract the body tends to shorten, thereby reducing its volume. However, as the body fluids are incompressible, the hydrostatic pressure in the worm increases. The fibres are themselves inextensible but by moving across each other the two series of fibres allow some flexibility in the cuticle (Harris & Crofton, 1957). The overall shape of a nematode is usually cylindrical with pointed ends and the body organs are bathed in pseudocoelomic fluid; being incompressible, the fluid transmits the pressure exerted on it equally throughout the body.

The remarkable uniformity in nematode structure may be correlated with the hydrostatic skeleton and the complementary structure of the cuticle (Harris & Crofton, 1957; Lee, 1965).

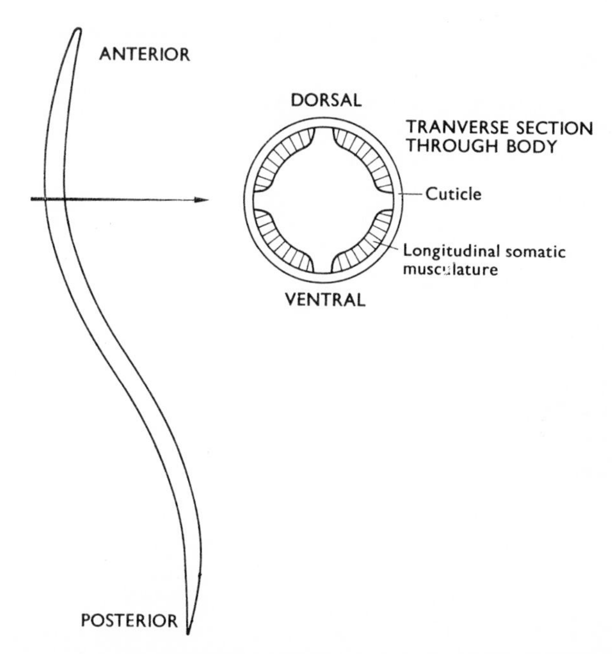

Figure 1 The arrangement of longitudinal somatic musculature for locomotion. The dorso-lateral and ventro-lateral muscle block pairs, oppose each other in contraction. The nematode therefore moves on its side, when moving in one plane.

The musculature

Nematodes are unusual in possessing no circular muscles for locomotion. Movement is by the alternate contraction of dorso-lateral and ventro-lateral muscles (Figure 1). These muscles generate a series of backwardly moving waves.

During normal wave formation the sub-dorsal and sub-ventral pairs of muscles in any section of the body are in opposite phases of contraction or relaxation. The waves, therefore, are formed in a dorso-ventral plane. Assuming that the anus is ventral and on a horizontal surface the nematode moves on its side (Figure 1).

Wave propagation

The combined effect of the hydrostatic pressure of the body fluids, the fibres in the elastic cuticle, and the longitudinal muscle bands, all in a flexible cylinder, provides a mechanism for propagating waves. Muscular contractions allow such changes in the overall shape of the nematode as determined by the cuticular structure and internal hydrostatic pressure. For further details see Harris and Croft (1957).

Oversimplification of the wave model

The model of forces and wave propagation outlined above is widely accepted, and all these forces are doubtless involved in nematode movement. Further study is required, however, to assess their relative significance in the locomotion of all nematodes. The marine nematode *Thoracostoma californicum* and the plant-parasite *Ditylenchus dipsaci* are both able to produce wave-like movements after being transversely cut into two pieces. Besides being able to move after the release of their internal hydrostatic pressure by cutting, nematodes can spiral and wave their anterior ends while fixed to the substrate by their tails. These movements depend on a complex co-ordinated contraction of muscles, beyond that allowed for by alternate dorso-ventral contraction alone. Wallace (1968) described the rotary action or helical movements of *Meloidogyne javanica* during penetration of root tissues and suggested that the right and left sub-dorsal muscles may function independently and in co-ordinated sequence, rather than just as dorsal and ventral muscles. Certainly the head ends of nematodes are more versatile than the somatic musculature.

The smaller nematodes which have been used for locomotion experiments have been plant parasites, animal-parasitic larvae, the laboratory-cultured vinegar eelworm, *Turbatrix aceti*, and *Panagrellus redivivus*, the sour paste eelworm. All are approximately 1 mm long, and when fully active their whole body is involved in wave propagation and conduction. However, larger nematodes like *Ascaris lumbricoides* or adult *Litomosoides carinii* and many others, although active, do not have the same co-ordinated wave propagation, and waves are not always seen passing directly along the whole body length.

Influence of a water film on movement

Wave-like or undulatory propulsion can only produce progression, as translocation from one place to another, if the waves can exert a force against external resistance. A snake unable to use its scales for purchase, moves forward in sand through the pressure it exerts at right angles to the granular particles. The longitudinal component of this force provides for forward progression. The direction of a snake's movement through sand may be ascertained by the 'piles' of sand pushed back during locomotion. Gray (1953) published a detailed theoretical treatise on undulatory propulsion, to which the reader is referred for further information.

Looss (1911) observed that the infective larvae of the human hook-worm, *Ancylostoma duodenale*, could only move over skin and penetrate when the skin was covered by a thin film of moisture. When the larvae were in greater depths of water they 'thrashed' ineffectively from side to side, being unable to progress without using the surface tension forces.

The propulsive force of an undulating organism swimming or crawling in water is derived from the resistance encountered by each element of the body as it is displaced in a direction normal to its own surface. Provided the coefficient of resistance to such displacement is greater than that to displacement tangentially to its surface, a forward movement results.

Many small nematodes, including free-living soil forms, animal-parasitic larvae and plant-parasites, use the surface tension forces of water films to provide the purchase necessary for progression. The surface tension force acting on a nematode in a water film has been estimated as 10^4 to 10^5 times greater than gravity (Crofton, 1954). In a series of experiments Wallace analysed the forces at work when a plant-parasitic nematode crawls with the aid of a water film. The principles he elucidated probably apply to all nematodes that crawl in aqueous films. In addition to water films nematodes gain purchase from solid particles like sand grains and hair, even when water films are excessive and the worms would otherwise 'slip'.

From the figure (Figure 2) it may be seen that the thickness and therefore the surface tension forces of the water films are critical in providing the forces for, and governing the form of progression. The surface forces act along the whole length of the worm, but it is convenient to consider the forces at one point, by examining a cross-sectional view of the worm (Figure 3, Wallace, 1959b).

The forces acting on a nematode at rest in a thin horizontal film of water may be summarized as: $2\mathrm{T} \cos \theta_1 + mg$, acting at right angles to the substrate; where T is the surface tension of the film, m is the mass of the

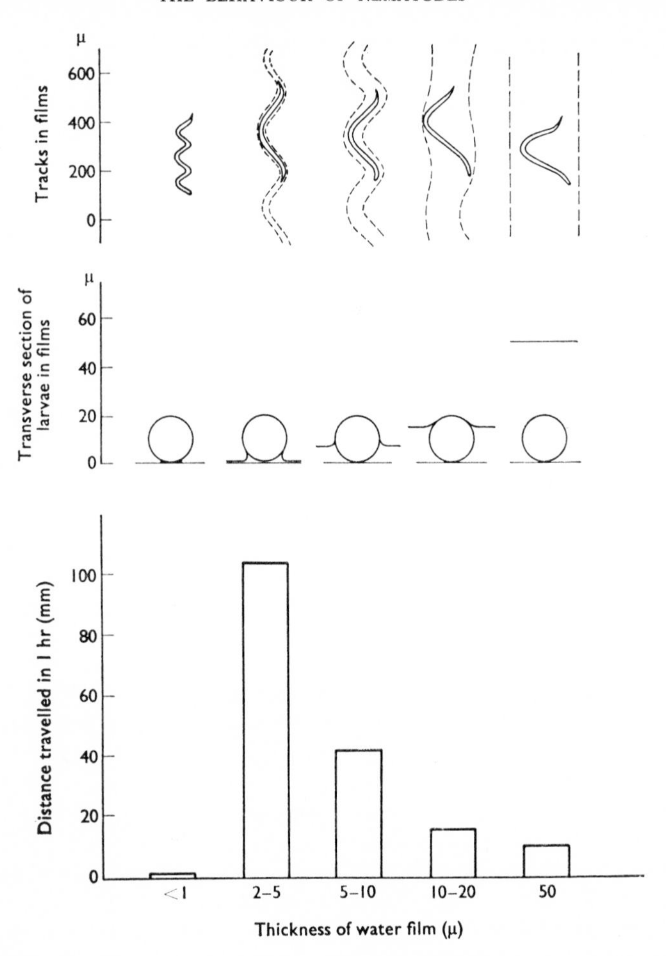

Figure 2 The relationship between the thickness of water films and the speed of larval movement in *Heterodera schachtii*. The top two graphs show the path taken by a larva in a film of increasing thickness. (Wallace, 1958a).

nematode, g is the acceleration due to gravity, and θ is the angle between the surface tension force and the vertical (Wallace, 1959b).

As the film gets thinner, the angle between the film and the vertical decreases, consequently the force 2T cos θ_1, pressing down on the worm, also increases (Figure 3). Sideways movement of the worm distorts the film so that the surface tension forces are displaced and oppose further sideways movement. Thus, the horizontal component R cos α of the resultant of the surface tension force, plus the friction force of the worm moving against the substrate F, provides the necessary reaction to the

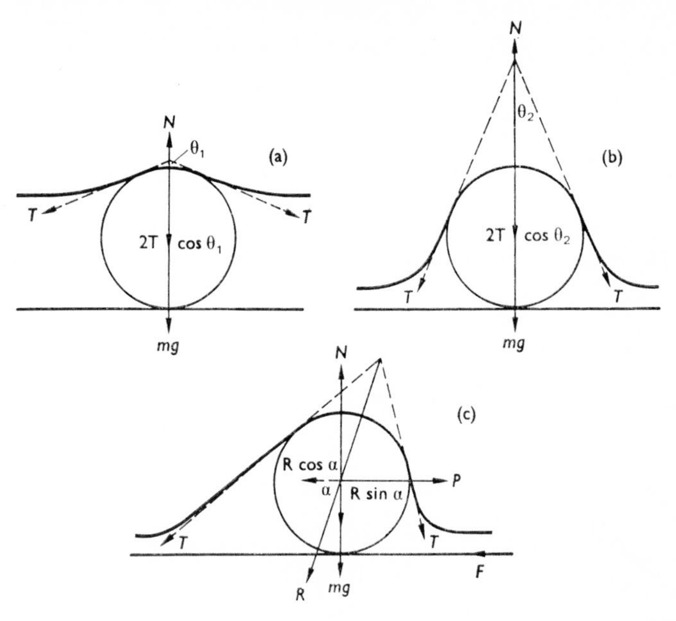

Figure 3 A model showing the forces acting on a nematode at rest (a and b), and moving (c) in a water film. Where T is the surface tension force; 2T cos. θ_1 the vertical component of the surface tension force; m the mass of the worm; and g the acceleration due to gravity; F is the friction force; and N the normal reaction to the surface tension force + mg. (Wallace, 1959b).

worm's own pressure on the film. The nematode therefore uses the film to generate a propulsive force, just as the snake uses sand particles.

If P is the force produced by the muscular contraction of the nematode, when $P = F + R \cos \alpha$ there is no lateral slip (Figure 3c).

The sinusoidal track made by a moving nematode in a water film is different if the thickness of the film varies (Figure 2). The wavelength is greatest and the amplitude highest in thick films, and as the thickness of the film approximates to the width of the worm the number of waves increases and the amplitude decreases. *Heterodera schachtii* larvae moved the greatest distances in unit time, when the surface film held the larvae firmly on the substrate, but not so firmly that the larvae could not overcome the friction forces.

The aqueous film acts as a rigid structure and the same forces are believed to act when a nematode pushes against soil particles for purchase. Much of this research was done by Wallace, and is reviewed in (Wallace, 1963 & 1968a).

The influence of surface tension forces on nematode locomotion has been so fully elucidated that the principles found are sometimes thought to be the same for all nematodes. The importance of surface forces for those nematodes that live in the soil or crawl in water films cannot be underestimated. There are, however, many nematodes that can swim freely in depths of water considerably greater than their own diameters, these include: *Turbatrix aceti* (Peters, 1928), *Aphelenchoides ritzema-bosi* (Wallace, 1959b), *Chromadorina viridis* (Croll, 1967), and a number of others including free-living stages of animal-parasitic larvae. The adult female of *Mermis subnigrescens* can move independently of any water, although they do usually emerge from the soil in damp weather (page 38).

Analysis of gliding and free swimming

The forces in gliding and freely-swimming nematodes have recently been analysed. Using microcinematographic techniques and analysing individual frames, Gray and Lissmann (1964) attempted to find out to what extent the form and frequency of waves passing down the body were determined by the nature of the external medium, and how the speed of a nematode was related to the form and frequency of the waves. They experimented with the following nematodes that glide: *Panagrellus silusiae, Haemonchus contortus* larvae, *Rhabditis* sp., *Strongylus* sp. larvae; and *Turbatrix aceti* which swims freely in vinegar.

Analysis of 'slip'

Under optimum conditions for progression, a nematode follows a path equal to its own width, without slip, and the rate at which the waves of contractions pass back is equal to the speed of progression. When there is slip, the waves are no longer stationary relative to the substrate but move backwards, this component of motion being Vn. If the waves move back relative to the substrate at a speed Vs, the speed of the nematode's progression (Vx) is Vw—Vs, where Vw is the speed of the waves relative to the head, and the percentage slip relative to the ground is 100 (1—Vx/Vw). The wavelength of the track left behind a moving nematode (λt), relative to that of the waves passing down it (λw), is determined by the amount of slip:

$$\frac{\lambda w}{\lambda t} = \frac{Vw - Vs}{\lambda w} = \frac{Vx}{Vw}$$

In studying the movement of nematodes the following were measured: wavelength (λ), wave frequency (f), and the speed of conduction over the body (Vw). Gray and Lissmann (1964) showed that, although the absolute speed of progression is greater in a worm that swims freely than in a crawling worm, the relative distance moved forward during the period of a single wave is much less. They also concluded that the ratio of amplitude to wavelength is substantially the same in both crawling and swimming movements.

Swimming freely in aqueous media

When the successive positions of a moving *Turbatrix aceti* individual were examined from the frames of a cine film, the amplitude of waves increased as they passed posteriorly along the body. In one individual the amplitude of lateral displacement of the head was about 35 μ, whereas the amplitude at the tail was 125 μ. This is characteristic of an organism moving by undulatory propulsion that can propel itself without 'yawing' (Figure 4, Gray & Lissmann, 1964).

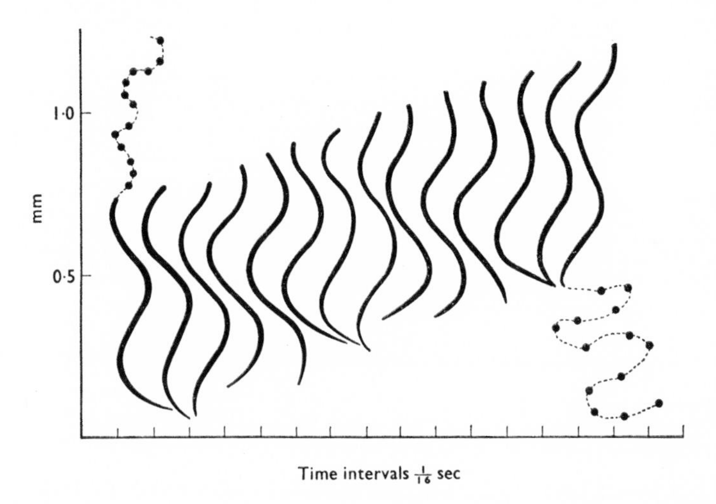

Figure 4 Successive positions in the swimming of an individual vinegar eelworm (*Turbatrix aceti*), at $\frac{1}{16}$ second intervals. The amplitude of the transverse movements of the tail was about four times that of the head. (redrawn after Gray & Lissmann, 1964).

Leaping

One adaptation in movement has recently been described for the infective larval stages of *Neoaplectana carpocapsi* DD 136 (Reed & Wallace, 1965). The larvae migrate to the surface of the substrate and move from one place to the next by 'bridging' between surface projections. When a bridge cannot be formed, however, the larvae leapt (Figure 5). A droplet was formed which held the larva to the substrate and another held the front end coiled against the middle of the body (Figure 5, a & b) holding the worm in a coiled position and preventing it from straightening out. When the larva overcame this resistance by muscular effort, it sprang free and 'took off' (Figure 5c, d & e). There was an optimum soil moisture content for the longest leap. Leaping resulted from the momentum developed by the front end.

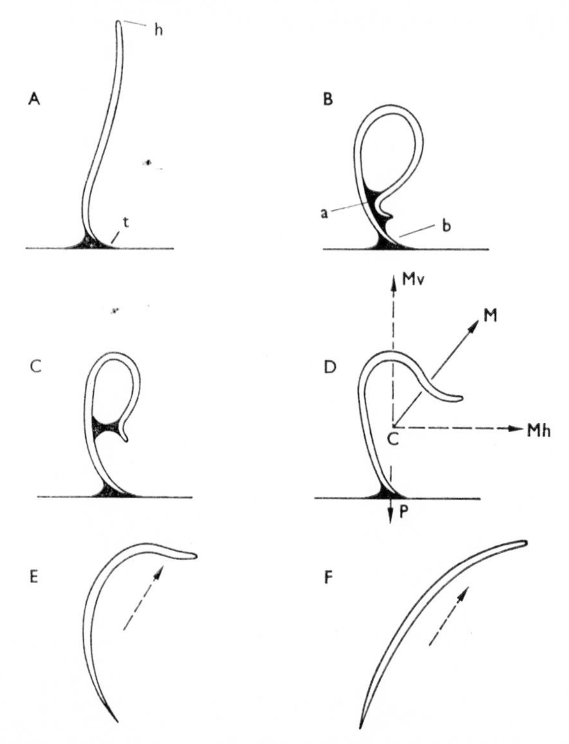

Figure 5 Leaping locomotion of *Neoaplectana carpocapsi* DD 136. A. Nematode vertical, attached by its tail (t), with its head (h) waving in the air. B. Nematode forms a loop, anterior point of adhesion (a), tail adhesion to soil (b). C Loop starts to open and the head swings upward. D. Restraining force of adhesion at tail (P) opposes vertical component (Mv) of momentum (M). Horizontal component (Mh), the centroid of the nematode (C). E. Mv being greater than P, the nematode 'takes off' and F. straightens out. (Reed & Wallace, 1965).

Movement within the egg

Larval *Meloidogyne javanica* move in the egg, and the forces needed have been analysed. Sinusoidal movement propels the larva, purchase being gained at the ends of the egg from the pressure acting at right angles to that exerted by the nematode. These forces resolve in a propulsive force and a displacement force. Undulatory movement inside the egg was through the longitudinal propulsive force and the frictional forces between the nematode and the egg shell. A steady glide occurred when they were equal (Wallace, personal communication).

Wallace (1968) thought there might be a correlation between movement in the egg, hatching, site of hatching and the shape of the egg.

Larvae in eggs which hatch outside the host may have to move before hatching, but this may be unnecessary in parasitic eggs that hatch within the host.

Other types of nematode movement

Most nematodes progress by undulatory movements and these either creep or swim, but others move differently.

Movement of *Criconemoides*

Besides sinusoidal movement, the soil dwelling plant-parasitic nematode *Criconemoides curvatum*, moves by alternate contractions and elongations following a pattern reminiscent to that in snails. The rear end contracts first and then a wave of contraction passes forward (Figure 6), only one such wave passing at any one time (Thomas, 1959). The movement of *C. curvatum* in water on a slide was recorded and analysed by micro-cinematographic techniques (Streu, *et al.*, 1961).

The cuticle of *C. curvatum* is composed of backwardly-directed annulations, and traction is provided by the retrorse edges of the annules. Contractions begin at the rear end and pass forwards, pushing the worm forwards. The superficial difference between this and the comparable movement of earthworms is that the contractions start at the tail and *push* the worm forwards, whereas in earthworms the head is first extended, anchored, and then the worm pulls itself forwards (Figure 6). One of the striking features of the film is that all the muscles in one section of the body contract simultaneously, which is entirely different from contractions in undulatory propulsion.

Streu *et al.* (1961) disagree with the earlier observations of Stauffer (1925) and Thomas (1959), who stated that the locomotion of *C. curvatum* was by an alternate lengthening and shortening of the body. After seeing the film the writer agrees with Streu *et al.*

Desmoscolex has a series of long cuticular setae, or bristles projecting from the rows of cuticular annulations. These bristles are used for a kind of 'walking' analogous to caterpillar movement. The waves are initiated at the rear end as in *Criconemoides curvatum*.

Another type of nematode movement occurs in *Chaetosoma* and is similar to that of a looper caterpillar (geometrid). *Chaetosoma drepanonema* has two rows of large hollow bristles, near the rear end. By alternately attaching and detaching its front and rear ends it can loop across the substrate. These movements appear to be specialized forms of undulatory propulsion.

Other movement

Work on the biophysics of nematode movement has perhaps overstated the mechanism of locomotion, while understating its biological function. Movement is essential in migrations to hosts or from dispoiled environments, but as Wallace (1958c) points out, many species move after food, and others, such as *Plectus* sp. remain anchored by their tail, waving their anterior end.

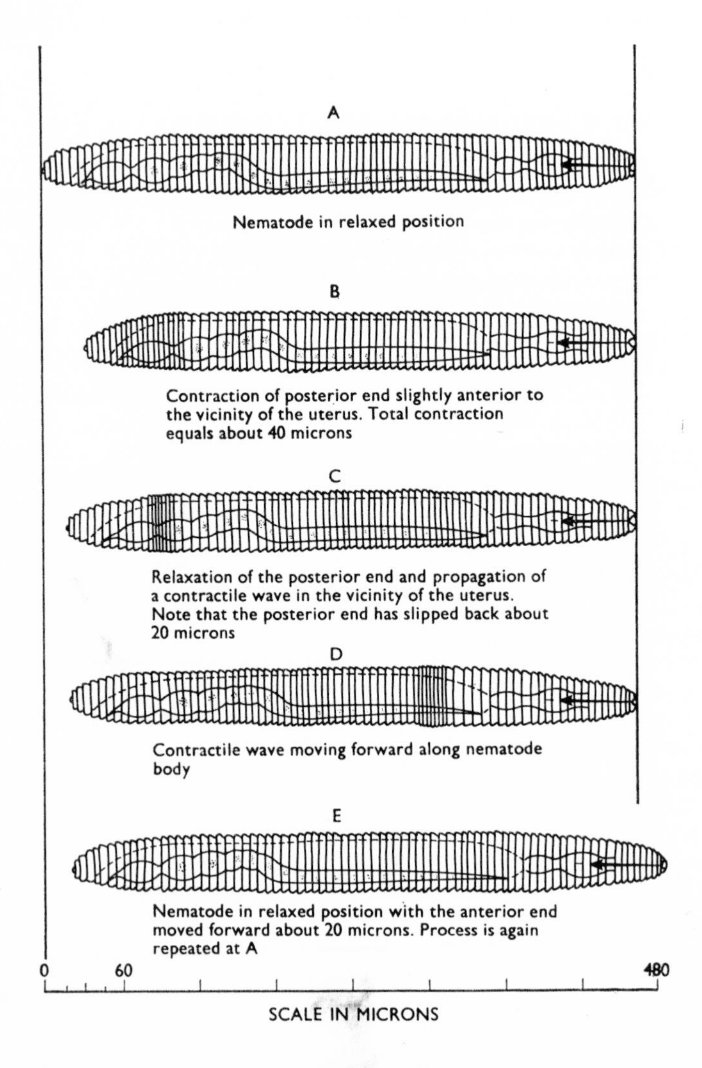

Figure 6 The peristaltic movements of *Criconemoides curvatum*. From A to E the wave of contraction which starts posteriorly may be seen to move forwards, as the nematode moves to the right of the page. (Streu *et al*, 1961).

Although most nematodes produce undulations, these may have become modified to serve a variety of non-locomotory functions. The first stage larvae of the Guinea worm, *Dracunculus medinensis*, flicks its tail in a way that resembles the movement of certain insect larvae. This mimicry is believed to enhance the possibility that it will be devoured by a copepod, the intermediate host of the Guinea worm. Mimetic behaviour is known in many other helminths (Croll, 1966d), and there are probably many examples in Nematoda not yet described.

3 | ACTIVITY, AGGREGATION AND SWARMING

When different species of nematode are observed their activity varies greatly. *Turbatrix aceti* is almost continuously active (Peters, 1928), as are the first stage larvae of the lungworm *Muellerius capillaris* (Rose, 1957), while *Dictyocaulus viviparus* infective larvae are mainly inactive (Robinson, 1962, page 96).

'Activity' is a nebulous term, open to different interpretations. Activity, by derivation and definition implies action, and in behaviour this action is usually movement. Differences in activity are, therefore, often described as differences in the rate or frequency of movement in unit time.

Inactivity and quiescence

Before considering nematode activity, the relationship between activity and inactivity will be considered. The ratio of time active to time inactive is an expression of the periodicity of activity. A sample of fifty adult *Pelodera* (probably *P. teres*), was kept in constant light and temperature for 6 days, and the percentage activity of the sample was estimated at regular intervals. All of the worms moved sometimes but a fairly constant proportion of 60 per cent (± 4 per cent) were active throughout the period. This led to the conclusion that *Pelodera* adults were active 60 per cent of the time. In the same way it was deduced that *Panagrellus redivivus* was active 78 per cent of the time and *Turbatrix aceti* 99 per cent. Diurnal and seasonal periodicities of activity are known from a large number of other invertebrate groups (Brown, 1965) and in the future may be demonstrated for nematodes.

Inactivity may result from the absence of sensory stimulation, sensory habituation, or the inability of worms to become active when stimulated, through a lack of available energy. Nematode inactivity may also be induced by: dehydration, cold, lack of oxygen and too high an osmotic pressure (Van Gundy, 1965b). Van Gundy attempted to standardize the terminology used to describe inactive states and followed closely the suggestions of Keilin (1959). Rejecting the term *anabiosis* as unsuitable, Van Gundy defines *dormancy* as a lowered but measurable rate of metabolism, *cryptobiosis* as a latent form of life where there is no apparent metabolism, and *diapause* as a temporary state of arrested development associated with a phase of morphogenesis. The best word describing most nematode inactivity is *quiescence*. Quiescence may be defined as a temporary, environmentally-induced inactive condition which ends when the environment again becomes favourable or includes suitable hosts or mates (Van Gundy, 1965a).

Inactivity may also be characterized by distinct postures or assumed shapes. When larvae of the human hookworm *Necator americanus* did not respond to stimulation, they were described by Payne (1923a) as being in a 'refractory condition'. In that condition the body was bent often in more than one plane, and remained so even when the larvae were tumbled over the bottom of the dish in a current of water. Khalil (1922) described a similar phenomenon in several other species, but was unable to observe it in *Haemonchus contortus* larvae, or any free-living forms. The 'kink' developed towards the tail of certain strongyle larvae when inactive (*Trichonema*, *Strongylus*), may also represent a refractory condition in these species. The early descriptions implied that the refractory state is a condition of recovery after activity, for it lasts only for a short period and is followed by normal sensitivity to stimulation. The physiology of this state is worth studying, for the maintenance of a contorted shape requires the expenditure of energy, working as it does against the hydrostatic skeleton.

Inactivity may also be the result of sensory habituation; *Trichonema* spp. third stage larvae did not respond to a light stimulus unless it followed a period of dark adaptation, although the same larvae responded to mechanical stimulation (Croll, 1966a).

An organism may become active through an endogenous internal stimulus, or in response to external stimulation (Figure 45); in both cases the response depends on the physiological state of the individual. The past history of an organism determines largely its present reactions, because it has also determined its physiological state. The major non-genetic variables are: ageing, starvation, the stage of development, the previous food source and the environment. For a discussion of some of these factors see Van Gundy (1965a).

Plant-parasitic nematodes are inactive for prolonged periods when the worms are usually described as 'quiescent'. *Aphelenchus compositae* and *Ditylenchus dipsaci* both tend to coil up when they are dried and the latter remains inactive in this state for a long time. The longest period of quiescence recorded is 39 years for *Tylenchus polyhypnus*, after which these phytoparasites recovered (Steiner & Albin, 1946).

The physiological basis for distinguishing between inactivity through sensory habituation, quiescence through unfavourable environmental conditions and the possible recovery of the refractory condition, is not understood, but sufficient information exists to suggest that the three states are not identical.

The biological significance of survival in an inactive state may be that the nematodes are enabled to inhabit environments not constantly favourable, and that a comparatively short life cycle can be expanded for prolonged periods (Van Gundy, 1965a). To estimate and discuss activity of nematode populations, without reference to the number inactive and the period inactive, may have little relevance in the field.

Measurement of activity

Most nematodes progress by sinusoidal propulsion which has led to the use of *waves per unit time* as a criterion in assessing the rate of nematode activity. When these measurements are made, one undulation is one complete cycle of events at the anterior end. Although this is a readily made measurement, comparison of measurements from different species may be unwise because of the great variation in the amplitude of the wavelength, which varies between species and is greatly influenced by the environment (Figure 2).

Another criterion commonly used for estimating activity has been to measure the *distance travelled in unit time* (Wallace, 1958c, and others). The drawback in this method is the assumption that the worms are travelling in a straight line, i.e. that the distance travelled is proportional to the distance moved between two points. If worms are placed in an intensity gradient, as is often done in behavioural studies, they may respond kinetically, and the distance travelled may have little relation to the shortest distance between the beginning and end. It must, however, be added that where both measurements have been made, there is some correlation between values expressed in waves per unit time, and those expressed as distance travelled (Table 1).

Table 1 The Activity of *Aphelenchoides ritzema-bosi* (after Wallace, 1960a)

Stage	Mean waves/min	Mean speed mm/min	Mean length μ
Adults	102	20	880
Intermediate larvae	46	5.1	550
Small larvae	16	0.5	230

Undulations themselves are only useful when they exert a pressure against soil particles, surface films, etc. (page 10), and therefore movement and activity cannot be directly equated with effective migration without reference to the environment.

Some nematode responses may be of the 'all-or-none' variety, that is, if there is a response it has a constant form and rate, irrespective of the intensity of stimulation. This kind of response is not recordable in waves per unit time or in distance travelled. The variable in this case is whether or not the nematode is responding, and not at what rate it is responding; such a measure of activity may be expressed as the *percentage or proportion of activity in a given population* (Wallace, 1962; Croll, 1966a).

When comparing the mobility of *Meloidogyne hapla* and *M. javanica*, Bird and Wallace (1965) measured, among other factors, the influence of temperature. For this, they used polythene tubing 2 × 5 cm, filled with sand of 125-250 μ particle size. After covering one end of the tubing with a nylon mesh, they placed the tube vertically in water. By placing aliquots of larval suspension on the upper surface, mobility was assessed as the number of nematodes that migrated through the sand and into the water in unit time.

Webster (1964) expressed a 'movement index' which was the average distance moved from the point of inoculation at one end of a tube towards the opposite end of the tube. Using a brass column 7·5 cm high divided into three 2·5 cm sections, *Ditylenchus dipsaci* were introduced at one end. Webster calculated the movement index in different conditions of temperature and suction pressure, using the following equation:

$$\frac{2\ L3+L2}{L1+L2+L3}$$

Where L1 is the layer into which the nematodes were introduced, and L2 and L3 were succeeding layers.

The activity of *Haemonchus contortus* and *Nippostrongylus brasiliensis* larvae was measured manometrically (Figure 12) by measuring the oxygen consumption before, during and after stimulation by light (Wilson, 1966). Manometric measurements monitor physiological processes more directly, but until more is known of the oxidative metabolism and the relationship between activity and oxygen consumption, it is difficult to compare these measurements with data on behaviour.

The relationship between nematode activity and oxygen consumption (QO_2) has been little studied (Nielsen, 1961). Only a small part of the oxygen consumed seems to be associated with movement (Von Brand, 1960). The QO_2 of infective *N. brasiliensis* larvae fell considerably after 12 days, Rogers (1948), and Santmeyer (1956) observed that the QO_2 of *Panagrellus redivivus* decreased to half its original value after 24 h starvation, but found mobility was unaffected.

Many other criteria have been used in attempts to measure the activity of nematodes, the choice usually being determined by the feature under investigation. Shepherd (1961) measured the rate of stylet probing during the hatching of *Heterodera schachtii* and Croll (1966b) the frequency of vulval contractions in *Mermis subnigrescens.*

All these measurements give only a partial picture of nematode activity; ideally the data relating to activity of populations should include at least the following values: undulations per unit time, percentage activity, and periodicity of activity. If an expression relating to distance covered is required, then the length of the worm and some information about the environment and the rate of turning should also be included.

Activity and the ecological niche of nematodes

Wallace and Doncaster (1964) compared the activity of nematodes of different habit and habitat: plant parasites of roots and of aerial parts of plants, larvae of animal parasites and free-living microbivorous forms. They reasoned that effective progression or velocity (V) was a product of length (L) and wave frequency (F). By plotting V against LF the nematodes studied fell into groups according to their habitat (Figure 7).

From this and other data free-swimming nematodes are apparently the most active and the next most active are the parasites of plant shoots and leaves, such as *Ditylenchus dipsaci* and *Aphelenchoides ritzema-bosi*. Following these come the animal parasites of grazing ruminants, which climb the

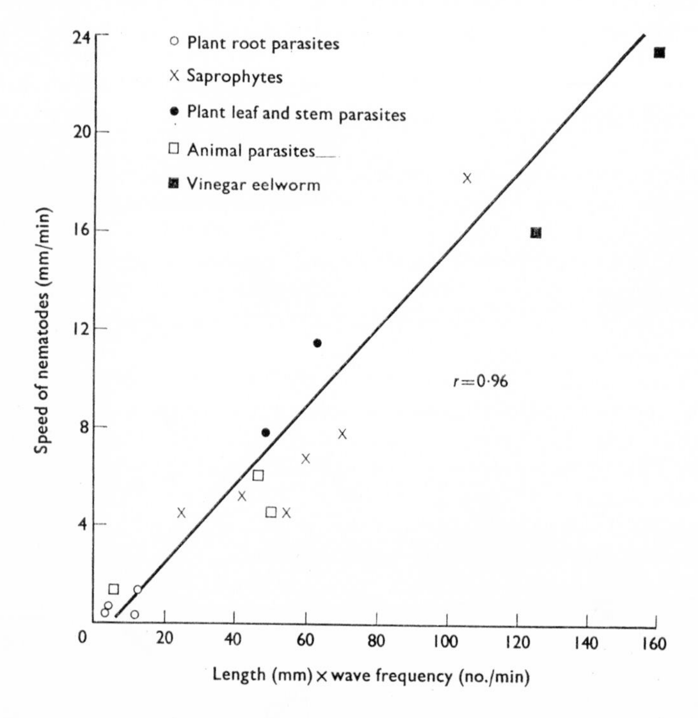

Figure 7 The relationship between the speed of nematodes and the product of their length and wave frequency, when moving in deep water. Each point is the mean for twenty worms. (Wallace & Doncaster, 1964).

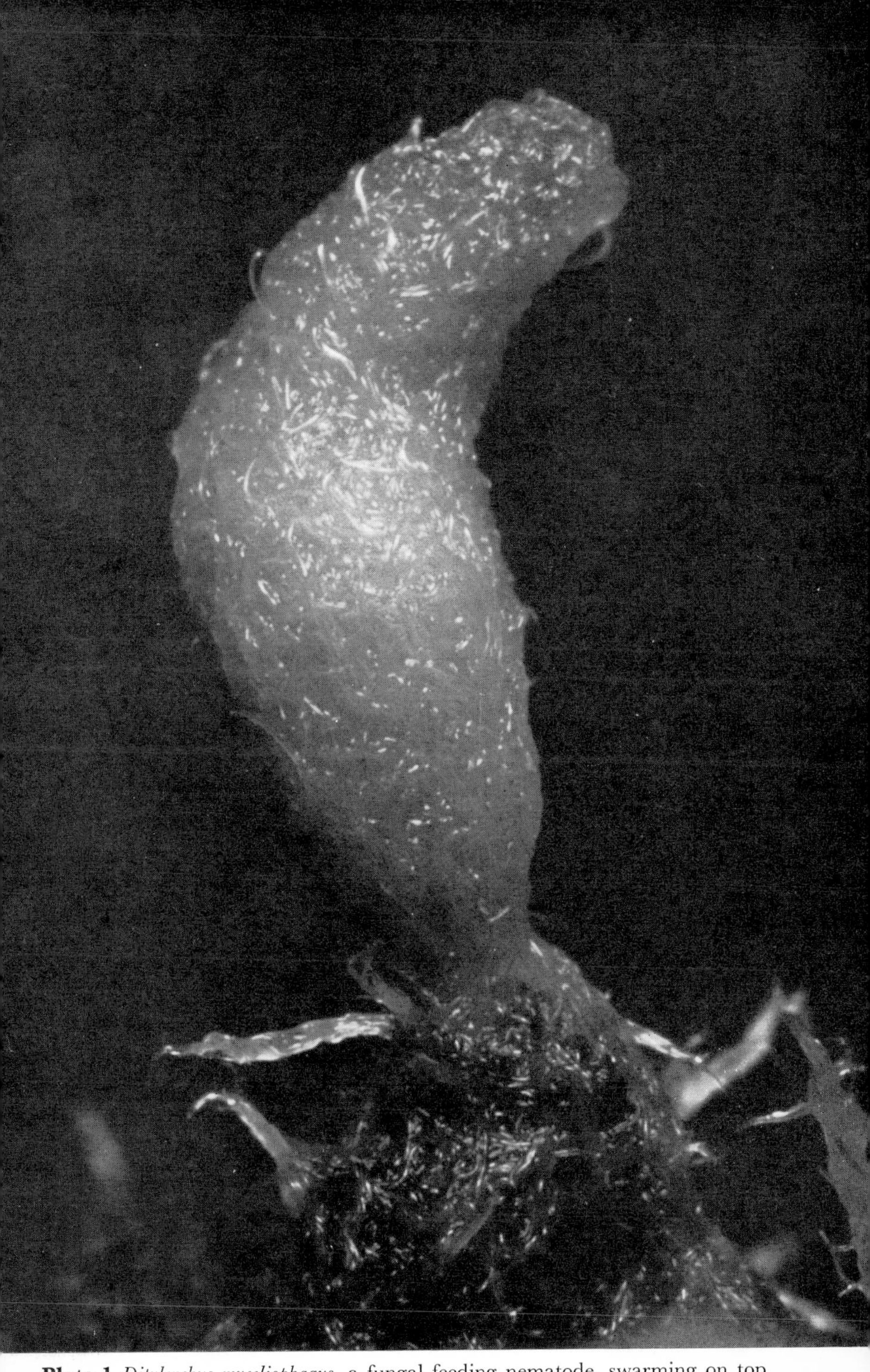

Plate 1 *Ditylenchus myceliophagus*, a fungal feeding nematode, swarming on top of the 'casing' layer of a mushroom bed. (Courtesy of J. J. Hesling, 1966.)

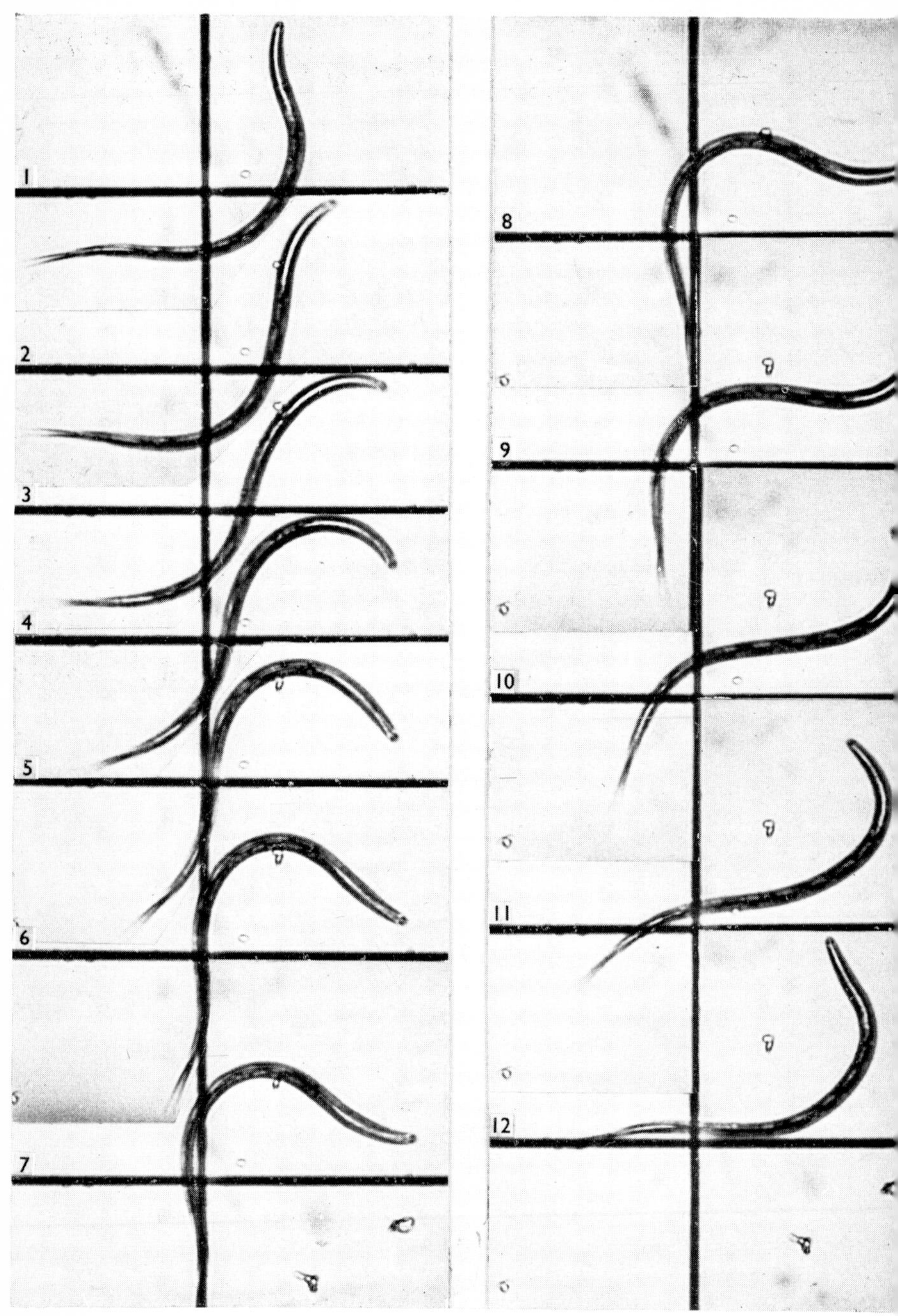

Plate 2 *Haemonchus contortus* larva in water, interval between photographs 1/22 second. following the position of the particle, and its displacement relative to the larva, the movem of the surrounding medium was observed during movement. (Courtesy of J. Gray & H. Lissmann, 1964.)

herbage to contact their hosts. The least active remain in the soil and include free-living forms and parasites of roots.

This postulate is based on a very few representatives from each group, and may be an oversimplification. The actively penetrating larvae of *Strongyloides stercoralis* are highly mobile; conversely, *Dictyocaulus viviparus* infective larvae are very inactive, although this may be correlated with its association with *Pilobolus* (page 96). *Pratylenchus penetrans* moved at speeds of 2 mm per minute in excised corn roots (Di Edwardo, 1960). Wallace and Doncaster's (1964) hypothesis that the activity of a nematode may be correlated with its ecological niche needs testing further, using a greater variety of nematodes.

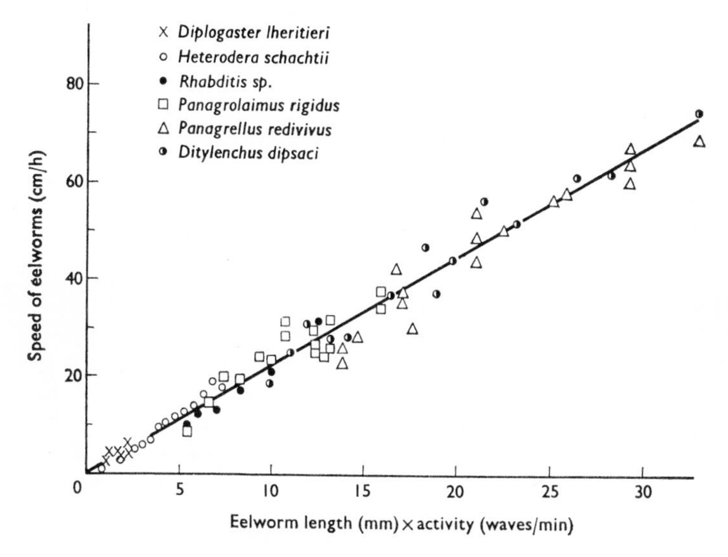

Figure 8 The relationship between nematode length, activity and speed among water droplets on a glass surface. (Wallace, 1958b).

Analyses of the tracks of six species of free-living and phytoparasitic nematodes indicated that the product of length and activity (waves per min) divided by speed is a constant (Figure 8). This applies only to movement in sand where the length of the nematode is less than three times the particle diameter (Wallace, 1958b).

Relationship between the stage or age and activity

The biological role of the different stages in nematode life cycles varies considerably. Some feed and grow, others are resistant, infective, migratory phases (page 2). Rogers (1962) emphasized that the infective stage is especially adapted to bridge the enormous gulf between free-living and the parasitic life. Bird (1967) demonstrated for plant parasites, the morphological and physiological changes which occurred in the infective second stage larvae of *Meloidogyne javanica* immediately following penetration of a host root.

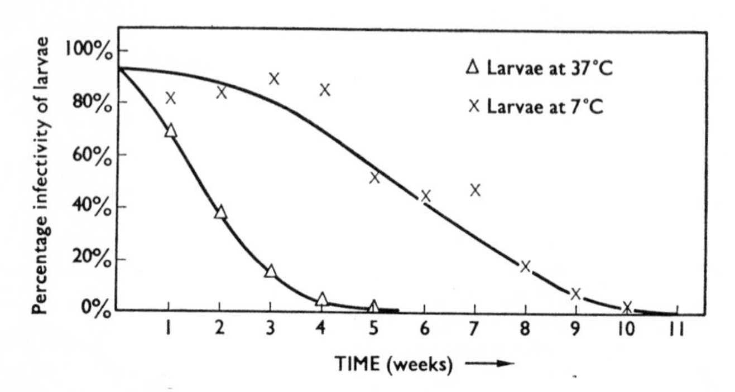

Figure 9 The changes in infectivity of hookworm larvae, *Ancylostoma caninum*, stored at 7°C and 37°C.

Payne (1922, 1923, and 1923a) showed that the activity of larval hookworms varied with age. The first and second stage larvae spent much of their time inactive, and moved more weakly than the third stage larvae. She concluded that: 'the young larvae must complete their feeding and reach the infective stage before the migration is begun' (Payne, 1923a). Food granules laid down in the first two larval stages were gradually reduced as the third stage larva aged. Rogers (1939 and 1940), studied the physiological ageing and activity of hookworm and *Haemonchus contortus* larvae further (Figures 9 and 10). The reduction of body contents

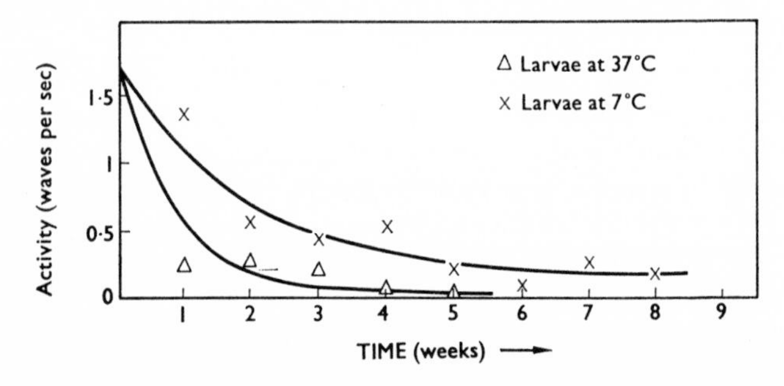

Figure 10 The changes in activity in *Ancylostoma caninum* larvae, stored at 7°C and 37°C. (redrawn after Rogers, 1939).

during ageing and starvation has also been studied in the larvae of *Meloidogyne javanica* and *Tylenchulus semipenetrans* (Van Gundy *et al.*, 1967) and found that the body contents were used up rapidly at high temperatures, in dry soils, and in oxygenated solutions.

These and other studies point to one conclusion: chronological age is not as meaningful, nor is it necessarily the same, as physiological age. Infective stages of both zoo-parasitic and phyto-parasitic nematodes have limited food reserves to provide the energy needed to locate and enter the host.

Under favourable environmental conditions these reserves usually allow enough time to reach the host, but under stress, or when the environment favours rapid metabolism, larvae 'age' rapidly.

After giving all the larval stages of *Trichonema* spp. the same light stimulus, Croll (1966a) observed that the first stage larvae were sluggish, moving for short periods only and then at wave frequencies rarely exceeding 10 per min; the second stage larvae were slightly more active, the periodicity of movement being infrequent. After the second moult, however, there was a steady increase in the wave frequency, reaching a peak between 10 and 20 days.

These and similar experiments demonstrate that there is a correlation between the activity, age and biological role of larval nematodes. Wherever comparative values are available, the free-living infective larvae are the most active. Observations such as these can be incorporated into generalizations on the relationship between activity, ecology and biology of nematodes.

Many of the extraction techniques used to separate nematodes from their substrate, including the familiar Baermann funnel, depend upon the activity of worms (Whitehead & Hemming, 1965). A greater knowledge of nematode activity would help in developing extraction techniques and in gaining some knowledge of the spread of nematodes in soil or on pasture.

Aggregation, swarming and synchronized movements

When suspended freely in an aqueous medium, some nematodes aggregate, forming densely matted clumps and clumping seems to be characteristic of some species or stages in the life cycle. Two plant-parasitic genera that show aggregation behaviour are *Ditylenchus* and *Aphelenchoides* both of which spend some of their life above ground level. There may be a survival value in aggregation through desiccation resistance or protection from ultra-violet light (Wallace, 1963).

Among the marine nematodes, *Enoplus communis* forms dense clumps when grains are lacking (Croll, 1966), as does *Thoracostoma californicum*. These responses are thought to be thigmokinetic (page 82). Rhabditid larvae extracted from week-old horse dung formed dense rosettes in which up to a hundred larvae were joined by the tips of their tails. Yoeli (1957) described a similar aggregation in microfilariae of *Wuchereria bancrofti* (plate 6).

The related phenomenon of 'swarming', or co-ordinated population movements, was recorded from the following genera: *Rotylenchus*, *Tylenchorhynchus*, *Hemicycliophora*, *Dorylaimus*, *Rhabditis*, *Ditylenchus*, and *Aphelenchoides* and it occurs in all sub-classes of Nematoda (McBride & Hollis, 1966). Infective strongyle larvae, *D. dipsaci* and *A. ritzema-bosi* and others swarm up the sides of culture vessels, but this may merely be a density dependent factor or be the result of surface film forces around the walls of the container.

One of the first descriptions of swarming in nematodes was that of *Rhabditis* sp. in mushroom beds exposed to light (Staniland, 1957). *Rhabditis* often occurs in millions on peat casing, 'the eelworms twisting round one another in glistening, writhing spires projecting from high points on the bed, easily visible to the naked eye' (Moreton, 1956). *Ditylenchus myceliophagus* and *Aphelenchoides composticola* both swarm from foul compost after they have destroyed the mycelium (Hesling, 1966; Carrol & Ritter, 1967). Hesling (1966) contrasts the swarming of these species with that of *Rhabditis* sp. (Staniland, 1957), the latter being more active and only swarming in the light, *D. myceliophagus* swarming in both light and dark. Hesling (1966) suggests that the swarming in *D. myceliophagus* on mushroom beds is an escape phenomen after the mycelium is totally destroyed.

The 'dauer' larvae of several nematodes associated with insects mount projecting object in cultures and in the field; these include *Neoaplectana carpocapsi* (DD 136) (Poinar, personal communication), and the dauer larvae of *Pelodera coarctata* phoretic on the dung beetle *Aphodius fimentarius*.

Hollis (1960, 1962) concluded that swarming in *Tylenchorhynchus* and *Hemicycliophora* spp. from soil was species specific, non-sexual and density dependent. That the properties of the cuticle are important in swarming was suggested when it was inhibited in *T. martini* by the endopeptidases papain, ficin and trypsin. These enzymes probably digest the proteinaceous layer of the cuticle.

Ibrahim (1967) also demonstrated the importance of the cuticle in swarming individuals of *T. martini* and found a difference in the external 'cortex' and other cuticular structures. He postulated that swarming was initiated by an internal mechanism, similar to that described for hatching and moulting (Rogers, 1960).

Swarming of *T. martini* after 60-75 days only occurred where host plants and all essential inorganic elements were provided (McBride & Hollis, 1966). They concluded that the difference between the swarming and non-swarming forms is that swarmers pull apart at a point of contact whereas non-swarmers can separate only by a gliding motion parallel to the plane of contact. Onset of swarming appears to be genetically controlled but determined by nutritional factors (Hollis, 1962) but this awaits confirmation, and represents only ideal responses.

Another phenomenon related to swarming, is 'synchronized movement' first described by Van Durme (1902) who later used the term 'petal' to describe masses of moving worms. Lane (1933) observed the same synchronized movement in hookworm larvae, and postulated that the larvae were negatively thigmokinetic (page 79): 'with larvae closely packed, such retraction of one larva will bring its opposite surface into contact with a second, which will shrink away at the spot touched'.

An example of synchronized movement may be observed in cultures of *Panagrellus redivivus* when the worms move up the walls of the culture vessel, but it is unknown whether this is an escape from harmful compounds or from the development of an unfavourable pH (Ellenby & Smith, 1966, Figure 27), or is a response to touch or gravity, or is the inevitable result of movement and surface tension forces (Lees 1953, Kämpfe 1963, and Gray & Lissmann, 1964).

Dean (personal communication) observed the aggregation behaviour of *Acrobeloides* n.sp. on agar plates. First the nematodes form a 'moving aggregation' where they all actively coil around each other and themselves. Next they form a 'bent aggregation' in which the worms are still active but most have a 180° bend in their bodies. Dean found that this stage was dispersed by light. Finally the worms form a 'spindle aggregation' where all lie parallel to one another. The spindle aggregations sometimes fuse and a few active worms migrate around the periphery. Hansen (personal communication) also observed dense aggregations in cultures of *Neoaplectana carpocapsi* and Lownsbery (personal communication) found aggregations in the agar under callus tissue cultures of *Pratylenchus vulnus*.

Aggregations, swarming and synchronized movements are widespread in the nematodes, and available evidence suggests that aggregation behaviour is associated with quiescence and resistance to adverse environments, whereas swarming is frequent in migration and dispersal.

4 | RESPONSES TO LIGHT

Evidence on the responses of nematodes to light is contradictory. Indifference, positive and negative phototaxes have been reported and heat from the infra-red portion of the incident light was not excluded in the earlier experiments which means that the apparent responses to light could have been responses to heat, or caused by convection currents in the water medium. Much of the earlier work also neglected to measure the intensity or wavelength of the light used.

Nevertheless responses to light have been described for many nematodes and reference to them cannot be omitted.

Summary of earlier studies

Positive phototaxes were recorded for the infective larvae of the human hookworm *Ancylostoma duodenale* by Payne (1922), and in the following year Cameron (1923) observed a similar response in *Bunostomum trigonocephalum*. Whereas *B. trigonocephalum* larvae were marked by positive phototaxis, *Haemonchus contortus* only moderate, and both *Heligmosomum muris* and *Ornithostrongylus douglasi* were negatively phototactic (Cameron, 1923). Sprent (1946) observed a movement towards light in *Bunostomum phlebotomum* larvae.

Negative phototaxes were attributed to the larvae of *Stephanurus dentatus* by Spindler (1934), *Dorylaimus* by Clapham (1931), and *Trichonema* larvae by Buckley (1940).

Peters (1928) first recorded indifference to light in the vinegar eelworm *Turbatrix aceti*. Buckley (1940) failed to demonstrate that *Haemonchus contortus* larvae responded to light but Cameron (1923) succeeded. Stewart and Douglas (1938) observed no photic response in *Trichostrongylus axei*. *Strongyloides stercoralis* were also indifferent to light (Fulleborn, 1924, 1932).

Cobb (1926, 1929) studied the effect of light on the gravid adult female of *Mermis subnigrescens*. This stage has a red spot at the head and which Cobb termed the 'chromatrope' (page 38). By holding different unspecified coloured filters between the sun and the worms, he found that the rate of uterine contraction was influenced by the wavelength of light falling on the chromatrope. Christie (1937) agreed that light was essential for oviposition in *M. subnigrescens*. .

Rogers (1940a) first measured light intensity in analysing the vertical migration of trichostrongyle larvae in the field, and although he made no attempt to isolate individual factors, he reasoned that light (sunlight) had influenced migration. The greatest migrations occurred at dawn and dusk, at light intensities of approximately 60 ft c (page 30). Nekipelova (1956) and others showed that trichostrongyles migrated at dawn, but found humidity important.

In experiments similar to those of Rogers, Rees (1950) studied the vertical migration of *Haemonchus contortus* larvae in experimental plots. Correlating the numbers of worms recovered on the grass with meteorological and micrometeorological factors, she concluded that light influenced migration, the greatest numbers being extracted from the grass occurring at low light intensities of 50-100 ft c.

Lees (1953, page 75) described *Panagrellus silusiae* as negatively 'phototropic' because it gathered on the dark side of the culture vessel, and climbed up the sides at night, and Staniland (1957, see page 26) described the swarming of *Rhabditis* sp. on illuminated mushroom beds.

The larvae of *Dictyocaulus viviparus*, the adults of which live in the lungs of cattle causing 'parasitic bronchitis' or 'husk', are carried from dung to the surrounding pasture by the fungus *Pilobolus Kleinii* (Robinson, 1962; Robinson, *et al.*, 1962). When illuminated the larvae are activated and climb up the erect hyphae onto the dark sporangium (see page 96). Soliman (1953) concluded that the first, second and third larval stages of *D. viviparus* were positively phototactic.

Working with the infective larvae of *Ancylostoma duodenale*, Aoki (1959) failed to demonstrate a positive movement towards light, but observed a negative phototaxis at temperatures over 31°C. This result opposed all previous work on hookworm larvae but suggested that the response to light may be dependent on temperature. Whether this is confirmed or not, it initiates studies of the interaction of stimuli in the behaviour of nematodes, a subject little investigated to date.

Following Cobb's observations on colour and Rogers' on light intensity, Parker and Haley (1960) did an experiment which further confused the study of nematode responses to light. They first demonstrated a positive phototaxis in larval *Nippostrongylus brasiliensis*, and then repeated their experiments including a heat filter. The negative results when using a heat filter led the authors to conclude that the apparent movement to light was in fact a movement to the heat emitted from the light source.

Wallace (1961) examined the effect of light on the behaviour of a plant-parasitic nematode, *Ditylenchus dipsaci*. Using a heat filter and a light gradient of 0-46 ft c, he concluded that light had little effect. Bloom (1964), however, observed that the chrysanthemum eelworm, *Aphelenchoides ritzema-bosi* migrated more readily in decreased light, possibly suggesting a form of kinetic response. Barraclough and French (1965) found the distribution of *A. ritzema-bosi* was uninfluenced by light, but only studied orientation responses.

In a review of the free-living stages of zoo-parasitic and phyto-parasitic nematodes, Wallace (1961a) concluded that there was little evidence that light influenced orientation at all. Between them, Parker and Haley (1960) and Wallace (1961, 1961a) had made suspect all of the previous experiments largely through the negative results they obtained when using heat filters. Wallace studied parastic nematodes, spending much of their lives in darkness, and Parker and Haley studied *N. brasiliensis*, which, being an active penetrator of warm-blooded hosts, may be more sensitive to heat than to light (page 48).

Unlike most of the above records, which were of free-living nematodes, or the free-living stages of parasites, Earl (1959) commented that both sexes of the filarial worm *Dirofilaria immitis* showed a sensitivity to visible and ultra-violet light when cultured *in vitro*. Upon exposure to bright light, adults of the filarial nematode *Litomosoides carinii* were invariably set into violent motion (Hawking, *et al.*, 1950).

Many of the above observations are directly opposed and no wide generalizations are possible; it may be noteworthy that of the thirty researches referred to above, six describe positive phototaxes, seven negative phototaxes, six indifference to light, and eleven suggest something but not a taxis.

Photokineses

A moving nematode may travel forwards or backwards, it may turn to the left or right at differing rates, or it may change its rate of activity, but in discussing kinetic responses it is with the last two of these alternatives that we are primarily concerned. Behind many of the observations outlined above there were suggestions that nematodes may respond to light in some way other than directionally.

Photo-orthokineses

An orthokinetic response to light was suggested when *Ostertagia circumcincta* was activated upon illumination following periods of darkness (Morgan, 1928). Croll (1966, 1966a) compared the orthokinetic activation of *Trichonema* spp. larvae with other nematodes. Following 24 h of darkness the larvae were exposed to a 'cold' light of constant intensity and spectral composition, and their activity estimated at 30 min intervals (Figure 11). *Trichonema* spp. infective larvae were inactive before illumination but quickly reached a peak of activity, then activity gradually decreased, until there was no movement after about 6-7 h. In continuous and constant illumination there was no further movement for 4 days, when the experi-

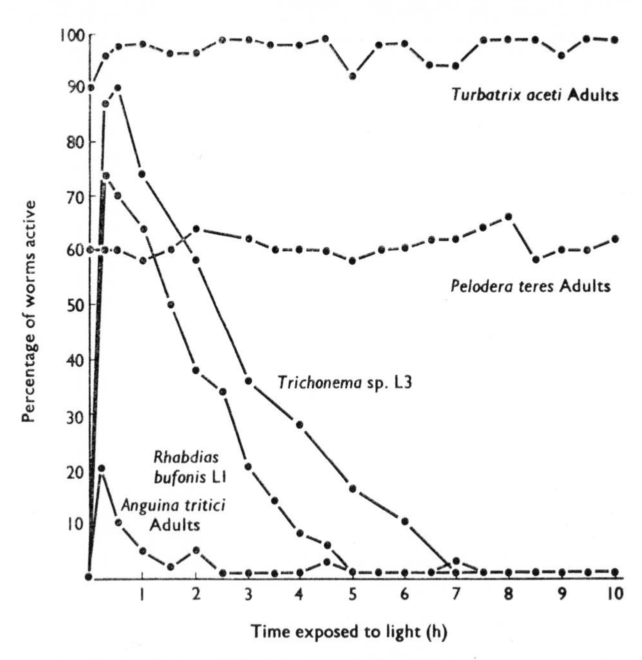

Figure 11 The photo-activation of different nematodes upon illumination, following 24 hours dark adaptation. (Croll, 1966).

ment ended. Figure 11 also includes results for *Pelodera* (probably *P. teres*) adults, mixed stages of *Turbatrix aceti*, *Anguina agrostis* larvae and the first stage larvae of *Rhabdias bufonis*. *Cooperia* spp., *Ostertagia* spp., *Trichostrongylus* spp. and *Bunostomum* spp. were activated upon illumination (Viglierchio & Croll, 1969).

These observations, however, differ from those of Wilson (1966) who used *Haemonchus contortus* and *Nippostrongylus brasiliensis* infective larvae. When stimulated by light the larvae of *H. contortus* increased their oxygen uptake at a greater rate than larvae in darkness, and when plotted (Figure 12) the QO_2, after an initial rise was nearly linear for almost 4 h, showing no decrease. The differences between these results may not reflect a difference in behaviour, but be the result of differential oxygen consumption for the QO_2 may not be directly related to activity (page 48).

Photo-klinokineses

'Infective larvae of *Strongyloides ransomi* avoid artificial light, but congregate in diffuse daylight' (Lucker, 1936); this may have been a klinokinetic response. Using infective larvae of *Trichostrongylus retortaeformis*, Crofton (1948a) found concentrations of larvae in the shaded area of his

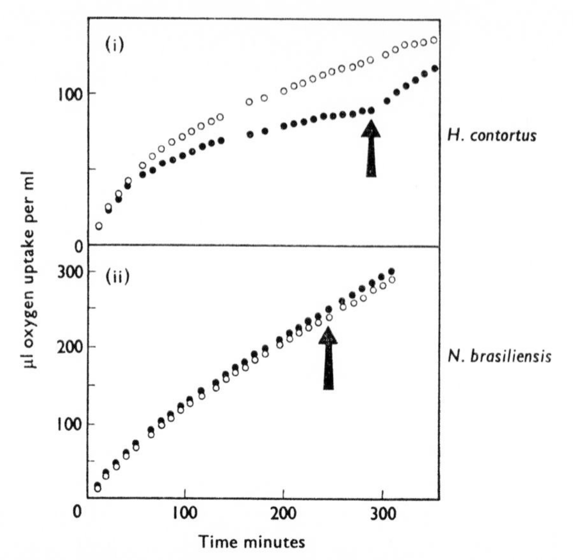

Figure 12 The oxygen consumption of (i) *Haemonchus contortus* and (ii) *Nippostrongylus brasiliensis* larvae, upon illumination. 0—0 larvae in the dark, 0—0 larvae in light 3200 lu./m². The arrows indicate where larvae in darkness were exposed to light. (redrawn after Wilson, 1966).

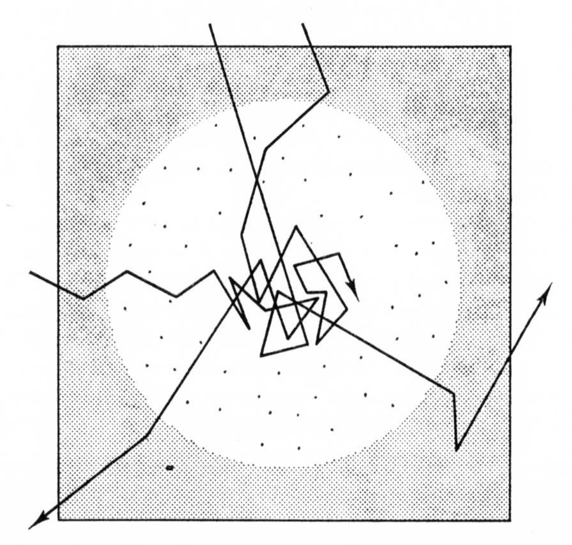

Figure 13 The photo-klinokinesis or change in the rate of turning of infective *Trichonema* spp. larvae, seen in the tracks of individual larvae across an illuminated arena. Heavy stippling, 5 ft.c.; light stippling, 300 ft.c.; and clear 900 ft.c. (Croll, 1965).

experimental vessel. Infective *Trichonema* larvae behaved similarly when the successive positions of the larvae were plotted at intervals as they entered and crossed a 'light arena' (Croll 1965, Figure 13).

From these experiments *Trichonema* spp. changed direction most, or turned most in bright light (900 ft c), and the next greatest rate of turning was at very low light intensities (5 ft c). The larvae moved in the straightest tracks and therefore moved the greatest distance between two points at 300 ft c. These results may be correlated with the great numbers of trichostrongyle larvae recovered from grass at dawn and dusk (Rogers, 1940a); Bees, 1960); the larvae may have followed the most direct paths and therefore travelled furthest during these periods.

Pigment spots and ocelli

Chitwood & Chitwood (1950) state that pigment masses in the Aphasmidia (=Adenophorea) have been known since the time of Bastian. Pigment granules in the marine forms *Enoplus* and *Oncholaimus* were described as ocelli, but Rauther (1907) suggested they were accumulated excretory material, a suggestion supported by Schultz (1931). Using a series of solvolysis tests Croll (1966c), found the pigment spots in *Enoplus communis* were very insoluble. Their solubility in strong alkali and in strong oxidizing agents suggested that they were a type of melanin. Such phenolic compounds are frequently associated with photoreceptors as linings that aid directional sensitivity. No evidence was found in *E. communis* to support this idea, and the granular pigment of the spot extended diffusely throughout the oesophageal musculature.

Schultz (1931) distinguished between pigment granules and true ocelli with associated lenses, but it is unknown whether they are photoreceptive organelles. The pigment spots of *Chromadorina viridis* were thought to be photosensitive (Croll, 1966e), and Wieser (1959), without evidence, deduced that pigment spots in marine nematodes were light receptors.

Between 40 per cent and 65 per cent of the total littoral marine nematodes have true ocelli or ocellar pigment which include such genera as *Leptosomatum*, *Deontostoma* (Figure 14a), *Eurystomina*, *Parasymplocostoma*, *Phanoderma*, *Calyptronema* and others. Compact granules of ocellar pigment occur in several of the Cyatholaimidae and Chromadoridae (Wieser, 1959).

Relatively few species of nematodes in littoral sand have ocelli, similarly to interstitial marine turbellarians, archiannelids, copepods and gastrotrichs (Remane, 1933). Wieser (1959) attempted to correlate the distribution of ocellar pigment with the habitat of marine nematodes (Table 2). *Thoracostoma* has, what Wieser calls 'ocellar pigment' and occurs on littoral and sublittoral algae, while the closely related pigmentless species *Pseudocella* is most common in the soft sublittoral bottom. This correlation also exists in the closely related genera *Phanoderma* with ocelli, and

Alyncoides without them, which is found in sand. Wieser (1959) suggested that ocelli and ocellar pigment are most frequent in forms inhabiting littoral algae and almost absent in species from littoral and sublittoral sand.

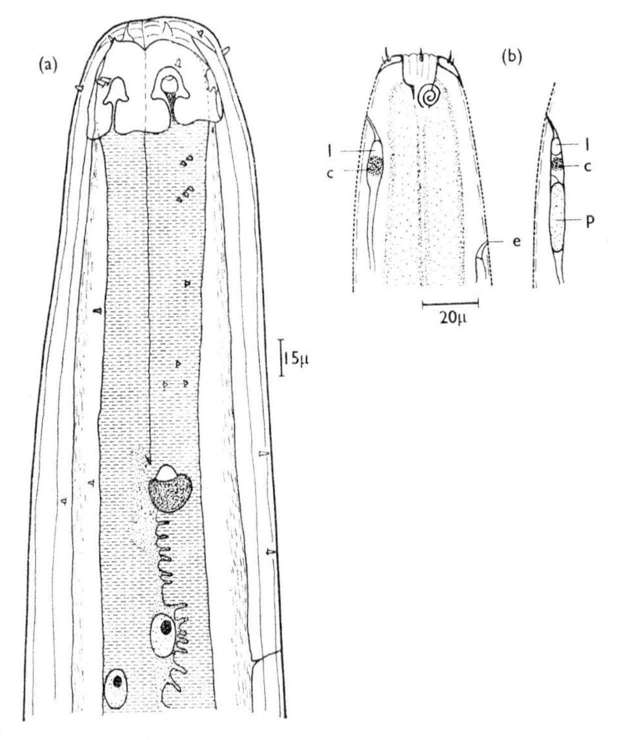

Figure 14 Ocelli in nematodes. a, The pigment spot of *Deontostoma magnificum.* (Timm, 1951). b, The ocellus of *Acanthonchus rostratus.* 1, the lensatic unit; c, the chromatic unit; p, the pigment cell; e, the excretory pore. (Murphy, 1963).

New genera and species have been made, using, together with othet characters, the presence or absence of ocelli, although it is known thar fixatives and clearing agents may dissolve or bleach the pigment. The presence or absence of ocellar pigment seems of doubtful taxonomic significance, unless observed in living specimens.

Little is known about the behaviour of nematodes possessing ocelli (that is, a lens and pigment spot), but Chitwood and Murphy (1964) observed that *Diplolaimella schneideri* was negatively phototatic They also reported that no photic response was observed for *Monohystera disjuncta.* The behavioural responses and sensory physiology of the nematode ocellus are almost unstudied, but might contribute to a knowledge of

Table 2 The Distribution of Marine Nematodes with Ocelli and Ocellar Pigment, in relation to their Habitat

	Littoral				Sublittoral		Littoral				Sublittoral	
	Algae exposed		Algae sheltered		Secondary substrates		Sand exposed		Sand sheltered		Coarse bottom	
	specimens	%	specimens	%	specimens	%	specimens	%	specimens	%	specimens	%
Ocelli	115	14·8	86	7·9	52	11·8	—	—	1	0·05	2	1·0
Ocellar pigment	378	48·7	243	22·4	34	7·7	2	0·2	—	—	1	0·5
Total	493	63·5	329	30·3	86	19·5	2	0·2	1	0·05	3	1·5
	%		%		%		%		%		%	
Ocelli	44·9		33·6		20·3		—		0·4		0·8	
Ocellar pigment	57·4		36·9		5·2		0·3		—		0·2	

[After Wieser, 1959]

photosensitivity in nematodes and to comparative invertebrate sensory physiology.

The colour of ocellar pigment and pigment spots varies widely, but most reports describe the pigment as orange or rust. Timm (1952) described *Monohystera microphthalma* as having no ocelli even in living specimens, while De Man originally observed distinct violet ocelli just behind the amphids. *Pontonema problematicum* has purple pigment spots (Chitwood 1960), but they disappeared in xylene. Both *Parachromadorella viridis* and *Nemella ocellata* have paired green pigment spots, the latter also having lenses. The red ocelli of *Eurystomina minutisculae* can survive formalin fixation, and one specimen of *Timmia* (=*Parachromadora*) *parva* was described as having one red and one green spot! This is an incomplete list of nematodes that have pigment spots but indicates the conflicting reports of pigment colour and solubility. Ocellar pigment or pigment spots are propably not all of the same chemical nature, some possibly being excretory products while others may be photo-sensitive.

Among nematodes, almost all of those with ocelli are marine. De Coninck (1965) considers the Areolaimidae of the infra-class Chromadoria are the most primitive group of nematodes. *Areolaimus elegans* has paired ocelli, and most other nematodes with ocelli are in the Chromadoria. Some fresh-water forms like *Chromadorina viridis* have pigment spots, but no soil-dwelling, phytoparasitic or zooparasitic nematode has ocellar pigment at any stage in the life history, except adult female *Mermis subnigrescens*. Therefore the possession of ocelli is primitive, and there has been a tendency to lose ocelli in all habitats except the marine. Some freshwater nematodes, that is, those with pigment spots, may have evolved directly from marine or estuarine types, whereas freshwater forms without pigmentation are more closely related to soil nematodes and have secondarily entered lakes and streams from soil.

Many descriptions of nematodes include reference to pigment spots. For example Timm (1951) stated that nothing was known of the innervation of the nematode ocellus, but refers to the 'bildungszelle' of Schultz (1931). This cell was believed to be the generative cell which secreted the ocellus in *Parasymplocostoma formosum*. Schultz also observed a canal, the 'augenkanal' opening from the lens to the exterior. Although Timm (1951) remarked that these observations were unconfirmed, they have been seen by Murphy (1963), in *Acanthonchus rostratus* (Figure 14b).

Timm (1951) described two large nuclei ($11\mu \times 18\mu$) a short distance behind the ocelli, embedded in the lateral walls of the oesophagus (Figure 14a) in *Thoracostoma magnificum*, *T. acephalatum* and *Leptosomatum elongatum*. These nuclei may produce pigment (Timm, 1951).

A possible phylogenetic relationship between nematodes and rotifers was postulated (Hyman, 1951). The similarity between the cerebral ocelli of rotifers and nematode ocelli is a small but additional piece of evidence supporting the phylogenetic affinity of the groups (Murphy, 1963).

The chemical basis of photosensitivity

The biological assumption underlying the mechanism of photosensitivity is that a pigment is present to absorb radiant light energy and translate it into a physiological stimulus. The pigment must absorb energy and be photochemically changed or activated to initiate a biochemical reaction. Of the photosensitive pigments known to zoology, carotenoids are the most widespread; in fact Wald (1946) advanced the thesis that, 'the photosensitive process depends almost universally upon one distinct group of substances—the carotenoids'.

Spectral sensitivity

The absorption characteristics of a pigment will be reflected in the response of the organism, as the response is mediated through the photolabile pigment. Consequently, the wavelengths of maximum absorption by the nematode pigment should produce the maximum response. Such a measurement of a behavioural response at different wavelengths of light is termed an 'action spectrum'. This is full of detailed difficulties, most of which have not been considered in nematodes, such as the refraction and spectral absorption of the cuticle and the numerous cellular inclusions through which the light must pass before reaching the pigment.

Action spectra are based on the absorption of quanta and in order to make valid quantitative comparisons of peaks, the intensity at each wavelength of incident light should be expressed as the number of quanta per second, per square centimeter (Seliger & McElroy, 1965). Such factors are further complicated in behavioural studies, e.g., the reflection of light as the organism turns. The incident photon spectral intensity distribution (photons/sec/sq cm) for a tungsten lamp of 120 ft c at 2800°K approximates to linearity between 400mμ and 700mμ, emitting ten times more quanta at the red end than the blue. For accurate construction of action spectra these considerations must be included for quantative comparisons.

By holding various colour filters between the sun and *Mermis subnigrescens* it was found that the rate of oviposition differed at different wavelengths. A reduced rate was observed in red and yellow light, while the rate of egg-laying was unaffected by Crooke's glass (filtering much of the near ultraviolet). Cobb (1929) concluded from these experiments that the gravid adult female of *M. subnigrescens* orientates itself and regulates oviposition by blue light. Christie (1937) and Croll (1966b) both found that oviposition was arrested or greatly reduced in the dark, and resumed on illumination. Adult female *M. subnigrescens* moved from red to green light if given the choice (Figure 15; Croll, 1966b).

Thirty-five years after Cobb's original description of the behaviour of *M. subnigrescens*, Ellenby (1964) found distinct *a* and *b* absorption bands from the chromatrope, and he concluded that the chromatrope

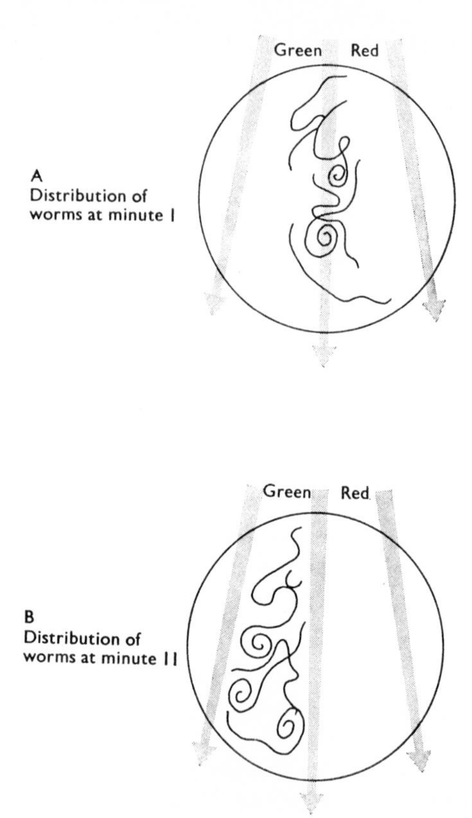

Figure 15 Colour preference in gravid adult female *Mermis subnigrescens*, given the choice between red and green light for ten minutes. (Croll, 1966b).

contained oxyhaemoglobn. Haemoglobin and its derivatives have never been shown to be involved in photosensitivity, although it is known that they are photolabile (Gibson & Ainsworth, 1953).

By illuminating the chromatrope of *M. subnigrescens* with different colours of light, and counting the rate of vulval contractions Croll (1966b) attempted to construct an action spectrum. The amount of data taken before the worms became moribund was limited, but the action spectrum suggested a peak of uterine contraction at 540 mμ, the *b* absorption band of oxyhaemoglobin (Figure 16a), it would be useful to repeat these experiments using more individuals.

The red pigment spots of *Chromadorina viridis* were found to disappear more quickly in light than in darkness after the nematodes were killed by heat or drying, suggesting that the pigment was photolabile. Using various solvents, the Pickworth benzidene test after heating the worms for

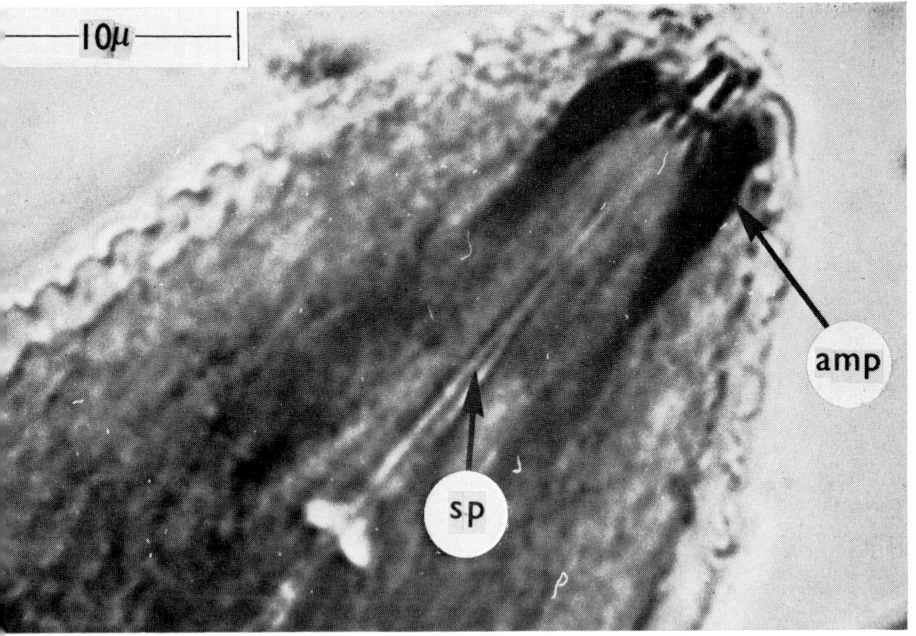

Plate 3 The head of *Meloidogyne*, stained for esterase activity, esterases may act in the sensory system. *amp.*, amphidial pouch; *sp.*, spear. × 3500. (Courtesy of A. F. Bird, 1966.)

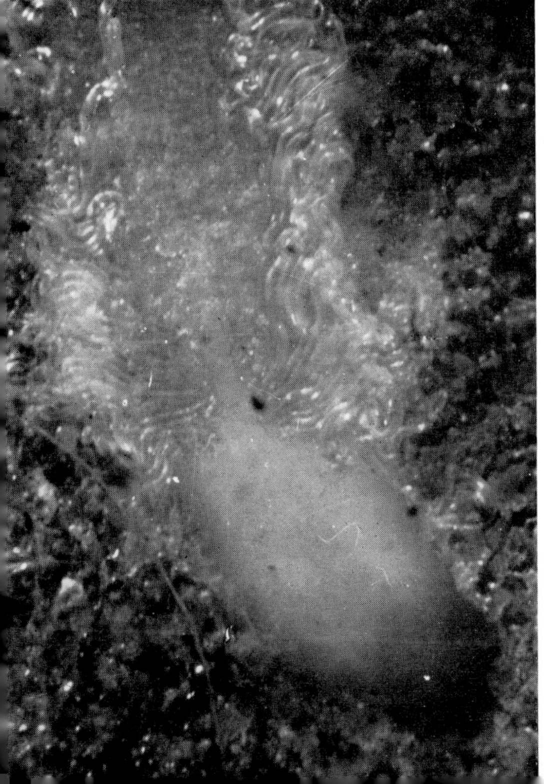

Plate 4 *Trichodorus viruliferus* massing on extending apple roots, dense numbers occur close behind the root tip. The root is about 1 mm. across. (Courtesy of R. S. Pitcher, 1967.)

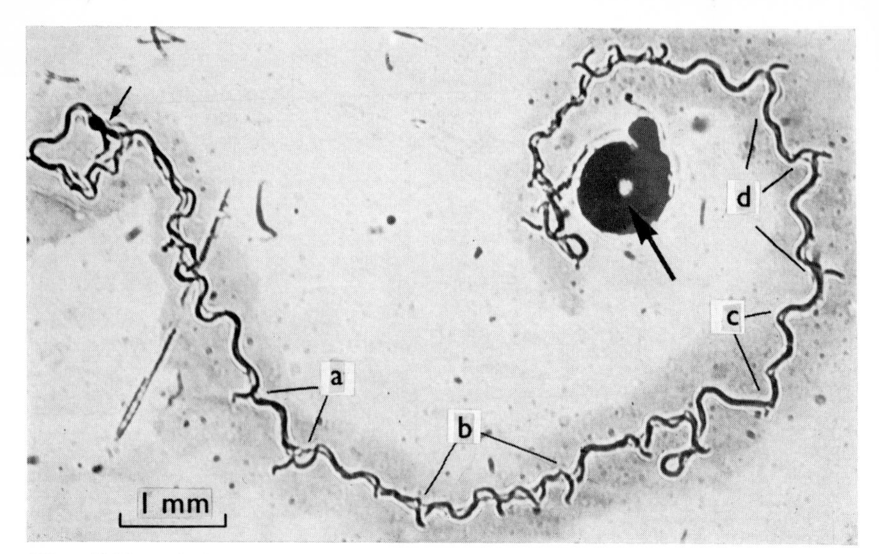

Plate 5 The spiral track of a male *Heterodera rostochiensis* approaching a female (large arrow), moving from the point of demarkation (small arrow). The track shows a number of patterns: (a) reversal, (b) reversal and turn after each wave, (c) indication of lateral probing, (d) less frequent reversal and turns. (Courtesy of C. D. Green, 1966.)

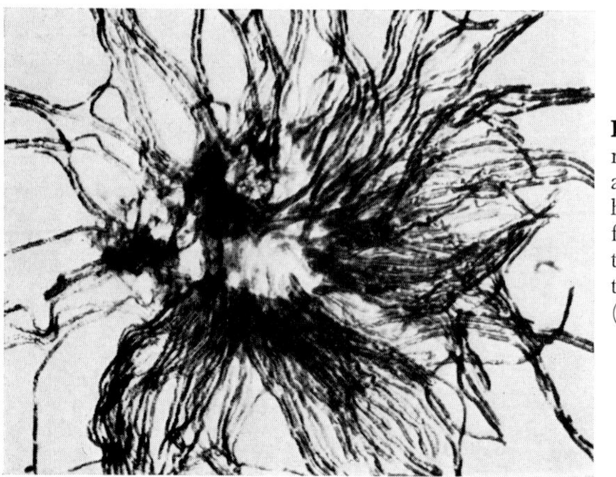

Plate 6 *Wuchereria bancrofti* microfilariae in human blood aggregated into 'medusa head formations'. The microfilariae are attached by the tail, and each clump contains up to 150 microfilariae. (Courtesy of M. Yoeli, 1957.)

denaturation of catalases and peroxidases (Lee & Smith, 1965), and constructing an action spectrum (Figure 16b) Croll (1966e) suggested that the pigment spots were a haem derivative, possibly haemoglobin or oxyhaemoglobin. Croll (1966e) also found a positive movement towards light of visible wavelengths shorter than 600 mμ, but a negative phototaxis at 366 mμ in the near ultraviolet. Whether these opposed responses are mediated through the same or different sensory receptors remains to be established.

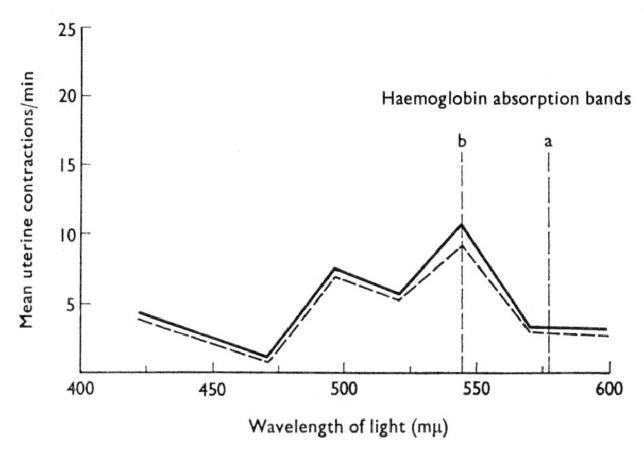

Figure 16a Action spectrum of uterine contractions per minute, as influenced by wavelength in *Mermis subnigrescens*. Dotted line represents one fresh individual, the continuous line is the mean for a number of worms. (Croll, 1966b).

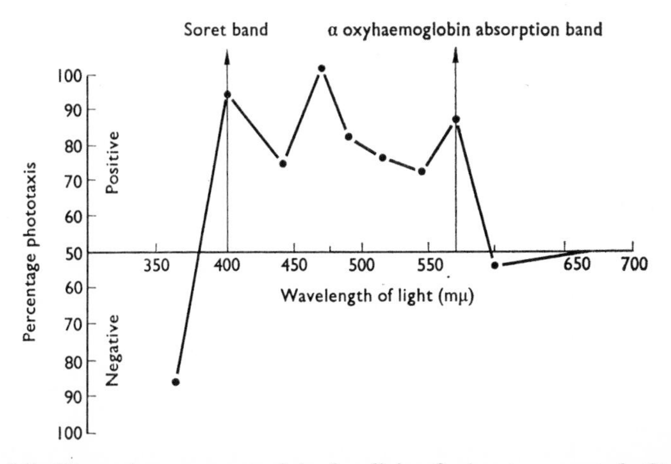

Figure 16b The action spectrum of the free-living freshwater nematode, *Chromadorina viridis*, showing the peaks of positive phototactic orientation in visible light, and the negative phototaxis in the near ultraviolet at 366 mμ. (Croll, 1966e).

Blue and ultraviolet light also stimulated adult *Dirofilaria immitis* to activity, in culture (Earl, 1959).

Light-dark adaptation

In his classical studies on the light sense, Hecht (1919) used the marine clam *Mya arenaria* and measured the reaction time for a photic response following increasing periods of darkness. In effect, therefore, he measured the rate of dark adaptation, and found that the reaction time (or time interval between stimulation and response) decreased with increasing dark adaptation. After a certain period he found that the reaction time remained constant, even in further periods of darkness. From his results Hecht argued that the reaction time was occupied by a chemical reaction, the velocity of which varied with the photochemical effect on the organism. The greater the photochemical effect the faster was the response, therefore, the shorter the reaction time. Hecht, presumed that something sensitive to light must have accumulated in the dark. In his own experiments he measured the length of time needed to elicit a response in one individual, with *Trichonema* spp. larvae; however, the length of time needed for complete adaptation was measured (Croll, 1966a).

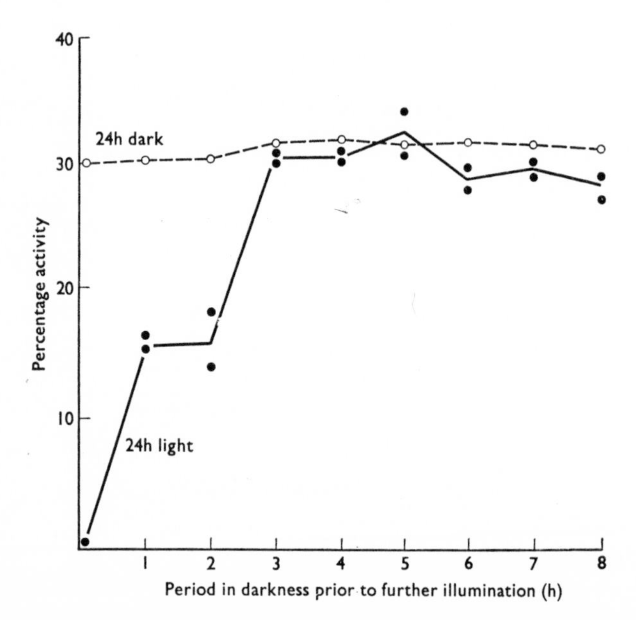

Figure 17 Dark adaptation in *Trichonema* sp. infective larvae. After 24 hours in light, batches of larvae were given increasing intervals of dark adaptation, i.e., 0 hr., 1 hr., etc. Open circles are controls, given 24 hrs. in darkness, then 25 hrs., 26 hrs., etc. (Croll, 1966a).

After exposure to light, batches of *Trichonema* spp. third stage larvae were given increasing intervals of dark adaptation before stimulation (Figure 17). It was found that at 20°C adaptation took about 3 h; adaptation did not occur in flashing light, but only in continuous light. It is not known whether the dark period is occupied by a bimolecular reaction, similar to that postulated by Hecht (1919), but if Hecht's symbols are adopted the following reaction may be postulated for *Trichonema* spp:

$$S \underset{\text{dark (slow)}}{\overset{\text{light (fast)}}{\rightleftharpoons}} P+A$$

A photolabile substance S is synthesized from two precursors P and A, the forward reaction is fast, and the reverse reaction is slow and is inhibited by light.

The dermal light sense

A dermal light sense (DLS) is very widespread in the animal kingdom, ranging through Protozoa (Jennings, 1906), Arthropoda (Kennedy, 1963) to man. A recent and comprehensive review on the DLS is that of Steven (1963), and it is with the characteristics of the DLS as he described them (Anonymous, 1969) that nematode photosensitivity is compared here.

Most nematodes for which a light sensitive response is known lack localized pigment or organelles and their responses may be mediated through a DLS. Steven draws a functional distinction between the optic and dermal responses, the DLS being associated with kinetic responses; removing the eyes of *Fonticola* for example did not affect its photoklinokinetic response.

The known spectral characteristsics of the DLS in other groups are not in close agreement with the spectral data for nematodes (page 37); furthermore, spectral sensitivities for nematodes have only been reported for those with localized pigment. The action spectra of the DLS in most animals have a sharp peak near 500 mμ, which falls off quickly as, for example, *Pholas*, *Mya*, *Cerianthus*, *Myxine* and ammocoete larvae. The lowest recorded peak is 455 mμ for the echinoderm *Diadema* (Yoshida & Millott, 1960). It may be significant that Viaud (1940), in his extensive studies on the 'sens dermatropique' of rotifers, gives a fairly broad action spectrum across the visible wavelengths of light. This suggests a closer relationship between the DLS of rotifers and nematodes than for the other zoological groups for which comparative data is available.

Evidence is accumulating from a number of invertebrates (Kennedy, 1963; Passano, 1963), to suggest that nerves may be directly sensitive to light. *Hydra* has no photoreceptors but highly developed photosensitivity, not responding to red light but responding strongly to blue light. The spectral features of *Hydra* photosensitivity are superficially similar to those of some nematodes. Pigmentless nematodes, being transparent, may well lend themselves to such a mechanism of dermal light sensitivity through photosensitive nerves or nerve endings.

Bionomics of responses to light

Light may act as a token stimulus, facilitating spatial orientation, whereby a nematode can distinguish between up and down. Such spatial orientation may have favoured the development of a phototaxes in *Chromadorina viridis*, leading it to areas of algal growth in a pond. Orientation to light may also have been selected for the gravid adult female of *M. subnigrescens* leading to oviposition near the grasshopper host.

In addition to direction, daylight is further characterized by its periodicity. It could be postulated that the orthokinetic response of *Trichonema* larvae upon exposure to light (page 30) may lead to activation at dawn when the light is dim and the humidity high, a period favourable for migration onto grass. Dark adaptation could then occur through the night, and on very bright days the influence of a photoklinokineses would tend to prevent extensive translocation into exposed areas (Croll, 1965, 1966).

The ultra-violet emission of direct sunlight is known to be lethal to many nematodes: *Rhabditis* was killed by light (Clapham, 1931), sunlight has been found to be lethal to larval *Heterodera rostochiensis* (Godfrey & Hoshino, 1933), and direct sunlight killed *Muellerius capillaris* first stage larvae (Rose, 1957). As well as providing nematodes with information about their environment, exposure to light may introduce important hazards, and this may be responsible for the many negative phototaxes recorded. It has been stressed that in many experiments on responses to light, heat has been inadequately excluded. In the field, however, heat and light occur together in sunlight. With heat comes desiccation and the possibility of death.

In attempting to summarize the biological relationship between nematodes and light, it is clear that there is some kind of balance between the hazards and benefits of exposure to daylight. In some species the balance may be biased in favour of exposure, while in other species behavioural responses may have evolved which tend to avoid light.

5 | RESPONSES TO TEMPERATURE

When considering responses to temperature, a distinction must be made between a sensory behavioural response and the thermodynamic effect of increased temperature. Light and heat are forms of energy, and temperature is a measure of the energy level of heat. The energy level influences the rate of metabolic turnover, movement and activity. As a result, larvae become inactive more quickly when stored at a high temperature than lower ones (page 24). Although direct thermodynamic effects may appear as behavioural responses, they differ in origin from those mediated through sensory receptors, the former being the consequence of changed rate of metabolic turnover.

Heat may be transmitted in three ways: radiation, conduction and convection, and temperature changes may also be associated with evaporation. Evaporation effects may be of considerable significance to organisms the size of many nematodes. Radiant heat, or the infra-red part of the electromagnetic spectrum, extends from the far red, about 760 mμ, to wavelengths of about 0·1 mm. The inadequate separation of visible and infra-red has led to the confusion in the interpretation of many experiments with light (page 28). Convection currents are frequently strong enough to carry nematodes through the fluid medium used in many experiments, and such effects have been offered as an alternative explanation for a number of described thermotaxes.

Activity and temperature

In common with almost all poikilotherms, the rate of nematode activity varies with temperature, broadly increasing over the range 5°C to 40°C with increase in temperature.

When heated, *Ostertagia circumcincta* larva became increasingly active up to 25°C, showing a reduced activity at 30°C, and eventually coiled up and were inactive at 40°C (Morgan, 1928). Commenting on his data, Morgan stated that an optimum of 23°C was unexpected, as the infective larvae encountered temperatures in the region of 38°C on entering its host. The larvae moved away from temperatures higher than 25°C. This observation suggests that the larvae are behaviourally adapted to remain in or around the optimal temperature (page 49).

Much similar data exists for other animal parasitic larvae: for *Bunostomum trigonocephalum* (Cameron, 1923 and Hesse, 1923), *Graphidium strigosum*, *Nematodirus filicollis*, and *Trichostrongylus retortaeformis* and numerous others. Rogers (1939) stated that 'those worms (*Ancylostoma caninum*) stored at 37°C used up fat reserve far more quickly than those stored at 7°C, evidently the temperature stimulated the larvae to greater activity causing a rapid reduction in the food reserve (Figures 9 and 10).

The narcissus strain of *Ditylenchus dipsaci* was relatively inactive at 10°C and the rate of activity increased linearly to 25°C (Webster, 1964), although Croll (1967a) found that activity in the same nematode was dependent on the previous storage temperature. It has been concluded that most phytoparasitic nematodes become inactive between 5°C and 15°C; the thermal optimum usually lies between 15°C and 30°C, and the nematodes usually become inactive again between 30°C and 40°C (Wallace, 1963). The same temperature relations would probably hold true for free-living and animal-parasitic forms, but the optima for animal parasites may be a little higher, and for actively-penetrating forms of warm blooded hosts they are highest of all.

In addition to the direct assessment of activity, the degree of migration at different temperatures has frequently been investigated in bionomical studies.

The vertical migration of *Trichonema* spp. larvae was greater from 13°C to 14°C than at other temperatures (Figure 31, Buckley, 1940). A related observation was made by Rogers (1940a), who recovered greatest numbers of trichostrongyle larvae off grass at 5°C and 45°C, while concluding that activity was minimal at 25°C. The data from these workers indicated a tendency, although neither set of results could stand a statistical analysis (Crofton, 1954). It was argued that there was no evidence for a reduction in activity between 13°C and 25°C because it contradicted both general observations and general principle (Crofton, 1954). In the light of Newell's (1966) studies on temperature in poikilotherms, this apparent anomaly is worthy of renewed attention.

There is little doubt that temperature affects the activity of nematodes, although many of the details are unknown. In addition to observations on the rate of activity, Bradley (1961) found that temperature affected the stimulation of a nerve-muscle preparation of *Phocanema decipiens*.

Temperature relations and distribution

The temperature optima and high lethal temperatures in a single species can vary between individuals collected from different geographical localities. Thus *Limulus*, the king crab, collected at Wood's Hole, Massachusetts, died at 41°C, while those from southern Florida about 2000 miles nearer the equator died at 46·2°C. Prosser *et al.* (1950), reviewing these phenomena, stated that *Planaria* tested at 30°C for sensitivity to cyanide showed a greater sensitivity if reared at 20°C than if reared at 30°C; and that clones of *Daphnia* (with distinctly differing temperature optima) have been segregated.

All of these phenomena may be grouped under the single heading of 'acclimatization', but the basic mechanisms of the phenomena are not properly understood. In vertebrates there is evidence that acclimatization is linked with the endocrine system. The melting point of fats in animals with high body temperature is higher than that in cold-blooded animals. Another set of explanations, possibly more applicable to nematodes, is that acclimatization is linked to changes in enzyme kinetics. A wide plasticity in thermal effect on enzyme systems may be expected from such nematode stages as infective third stage larvae. In the field these stages survive at temperatures around 10°C, but within seconds of entry into their hosts they are at 35°C to 37°C.

From the evidence of comparative physiology, therefore, it would seem unlikely that figures of high lethal temperatures and optima for activity and development, factors often investigated, would be the same for all the members of a single species. More probably the temperature relations would be correlated with physiological acclimatization (Croll, 1967a), and geographical distribution of the worms.

To date, this potentially rewarding aspect of investigation has received little more than occasional attention. Bird and Wallace (1965) studied the influence of temperature on *Meloidogyne hapla* and *M. javanica*. Although there is a considerable amount of geographical overlap, *M. hapla* occurs more frequently in colder climates whereas *M. javanica* is more prevalent in hotter regions. From this premise it was argued that temperature may be a contributing factor in determining the geographical distribution or thermal optima of the two species.

Table 3 Approximate optimum temperatures for various stages in the life cycle of *Meloidogyne javanica* and *M. hapla* (Bird & Wallace, 1965)

	M. javanica	*M. hapla*
Hatch	30°C	25°C
Mobility	25°C	20°C
Invasion	—	15°C–20°C
Growth	25°C–30°C	20°C–25°C(?)

From these results Bird and Wallace (1965) concluded that in general *M. javanica* has a higher optimum temperature range than *M. hapla*. They also found that there were different thermal optima for different biological activities. As the authors point out, the ranges are sufficiently close to allow the two species to occur together, but at the climatic extremes of the range one or other would dominate. Similar studies, using a single species with a worldwide distribution, may disclose behavioural and physiological differences that could be correlated with geographical acclimatization.

Balasingam (1964) made a comparative study on the effects of temperature on the free-living stages of the three North American hookworms: *Necator stenocephala* in the wolf, *Placonotus lotoris* in the raccoon, and *Ancylostoma caninum* from the domestic dog. In nature the distribution of these three hosts and their respective parasites extends across the entire continent, the wolf being found in the Arctic circle, then further south the raccoon and finally the domestic dog which is found in tropical and subtropical areas. No behavioural studies were undertaken, but Balasingam (1964) was able to demonstrate a close correlation between the regional temperature and the optimum thermal conditions for development, corresponding to the geographical distribution of the three hookworms.

The types of influence that thermal acclimatization and geographical distribution are known to have on temperature relations, must inevitably make the comparison of data from different parts of the world, and of nematodes of varying physiological history, essentially speculative.

Influence of fluctuating temperatures

Cysts of *Heterodera rostochiensis* the potato-eelworm or golden nematode were given a variety of temperature treatments, and the number of emerged larvae was compared with cysts kept at constant temperatures, and with those exposed to temperature fluctuations (Figure 18, Bishop, 1955; 1955a). It was found that a fluctuating temperature accelerated

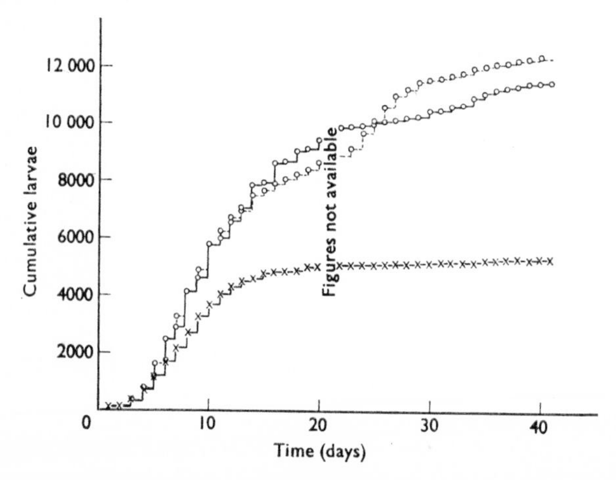

Figure 18 The influence of a fluctuating temperature on emergence of *Heterodera rostochiensis* larvae: cumulative curves showing the influence of temperature on emergence. X—X constant temperature, 0—0 alternation once in 24 hrs., 0—0 alternation once in 48 hrs. (Bishop, 1955a).

the rate of larval emergence from the cysts. Bishop related the phenomenon of an alternating temperature to the diurnal changes in soil temperature, quoting the range of fluctuation in a potato field as ranging from 9°C in early May to 15°C in mid-June.

A similar result of fluctuating temperature was found for *Heterodera schachtii* larvae, where the mean temperature applied to give optimal emergence was less than the optimum temperature for emergence (Wallace, 1955). No comparable increase in larval emergence occurred when cysts of *Heterodera glycines* (Slack & Hamblen, 1961), or *Meloidogyne javanica* egg sacs (Wallace, 1966), were exposed to conditions of alternating temperature.

Thermotaxes

Khalil's (1922) experiments on the thermotaxes of hookworm larvae were some of the first investigating nematode responses to heat. Khalil described the movement of hookworm larvae towards a heat source, a response that has been confirmed many times since:

'When a large number of larvae are used, the process can be watched with the naked eye as a wriggling opalescent mass which disperses gradually when the hot glass is removed and re-collects when it is reapplied. If the heated point is shifted to another place the mass of larvae follows it, producing a slightly curved column of active larvae streaming towards the new point of application.'

If the description of a *curved* mass lends suspicion that convection currents influence the response, Khalil's floating coverslip experiment may help to confute this criticism. A coverslip was freely floated in a glass cell filled with water, and touched with a heated rod, the larvae were observed swimming upward from the bottom towards the heated coverslip. Fuchsin particles were added and their distribution compared with that of the larvae. The particles were carried to the heated point, and were immediately carried past, but the larvae remained at the heated point, at first being carried by the current and then heading into it.

These responses occurred in the larvae of *Ancylostoma duodenale*, *A. ceylanicum*, *Necator americanus*, *Strongyloides stercoralis*, *Galoncus perniciousus* (from the tiger) and *Trichostrongylus douglassi* (from the ostrich). No such response was found in *Haemonchus contortus* larvae or in any free-living nematodes. Later work has shown that Khalil went too far in concluding that active-penetrating larvae responded to heat, whereas free-living nematodes and free-living stages of other nematodes did not.

Lane (1930) argued that worms swimming towards a heat source (on a purely random basis) moved more quickly and therefore tended to approach a heat source, and the worms that swam away moved more slowly. Lane believed that an active concentration of worms would inevitably accumulate around the heat source.

Fulleborn (1932)* emphasized the hazards of convection currents but he was able to conclusively demonstrate a positive thermotaxis in hookworm larvae on agar plates. Later, Lane* (1933) agreed that a positive thermotaxis did exist, and this observation, although never explained in terms of sensory physiology, has not been disputed since.

McCue and Thorson (1964) studied the behaviour of the parasitic stages of *Nippostrongylus brasiliensis* and various other adult helminths in a thermal gradient. They first placed adult *N. brasiliensis*, fresh from their hosts, into a gradient ranging from 28°C–48°C, and found all the worms at 48°C. Next they used a gradient of 37°C–52°C, where they found that all the worms had again moved to the hottest end and had died through thermal damage. They repeated these experiments several times with identical results and observed a strong positive thermotaxis, where the nematodes were unable to recognize temperatures lethal to them.

In 30 min female *N. brasiliensis* moved 23·6 cm over the 36°C–52°C gradient; McCue and Thorson calculated that if the thermal receptors were on opposite ends of the worm (an arrangement they admit to be unlikely), a worm 5 mm long would detect a change of 0·3°C. If the receptors were 1 mm apart the sensitivity would be to a temperature differential of 0·07°C. In a subsequent paper it was reported that the age and circadian rhythm of the rat host influenced the speed of the thermal response (McCue & Thorson, 1965).

When the third stage larvae of *N. brasiliensis* were subjected to a rise in temperature from 25°C–30°C or to 38°C, there was a rapid increase in the oxygen consumption (QO_2). This was followed by a decline in the QO_2 after 3 to 5 h. Wilson (1965) suggested that this was due to a physiological 'overshoot' and that the response may be independent of the particular rise for the range 30°C to 38°C, but that it is a response to the change mediated at a neuro-muscular level. Wilson (1965) claimed this to be a metabolic expression of the so-called 'thermotactic' response of skin-penetrating larvae of some strongyloid nematodes.

Suicidal thermotaxes have also been described for *Rhabdias bufonis* adults from frogs, adult *Camallanus* from turtles and *Haematoloechus* (digenean trematode from frogs) and *Neoechinorhynchus* (an acanthocephalan from turtles). To compare these results with a free-living form, McCue and Thorson (1964) repeated their experiments using the thermophilic enchytraeid *Enchytraeus albidus*, which moved up the heat gradient to 40°C and remained there.

The suicidal thermotaxes of these parasites compare with the behaviour of the parasitic itch mite *Sarcoptes scabeii*, which shows a lack of any avoidance reaction, moving into a heat gradient until it is killed by thermal damage. Thermal damage was also observed in the plerocercoids of *Spirometra mansonoides*, after it moved into a heat gradient (Thorson,

*The rivalry that existed between Lane and Fulleborn is particularly colourful and may be followed in their papers.

et al., 1964). Suicidal thermotaxes may be the result of a lack of negative feedback in the sensory system.

Pratylenchus penetrans, *Ditylenchus dipsaci* and *Tylenchorhynchus claytoni* moved up small heat gradients of 0·14°C to 0·56°C over 4 cm on agar plates, but *Trichodorus christiei* and *Xiphinema americanum* showed no responses over these gradients (El-Sherif & Mai, 1968; 1969). They concluded that temperature gradients of the same order as those to which the nematodes responded occur around growing roots and in germination. They also speculated that nematodes may respond to infra-red radiation of plants.

Most of the nematodes tested in a heat gradient orientate with respect to heat, and the majority are thermo-positive. There are cases, however, of nematodes, e.g., *Dorylaimus saprophilus* (Calpham, 1931), which move away from a heat source.

Eccritic temperature or thermal preferendum

Not all heat responses shown by nematodes are towards the hot or cold end of a thermal gradient. Ronald (1960) investigated the influence of temperature on the infective fourth stage larvae of *Terranova decipiens*, from the muscles of cod. Mobility, mortality and the thermal preferendum were studied with respect to temperature. The larvae moved towards the hot end of a thermal gradient until they reached 32·5°C, where they remained. Above 32·5°C the larvae moved down the heat gradient until they reached 32·5°C. They were also most active at this temperature, and survived longer at 32·5°C than at 31·5°C or at 37°C. (Figure 19).

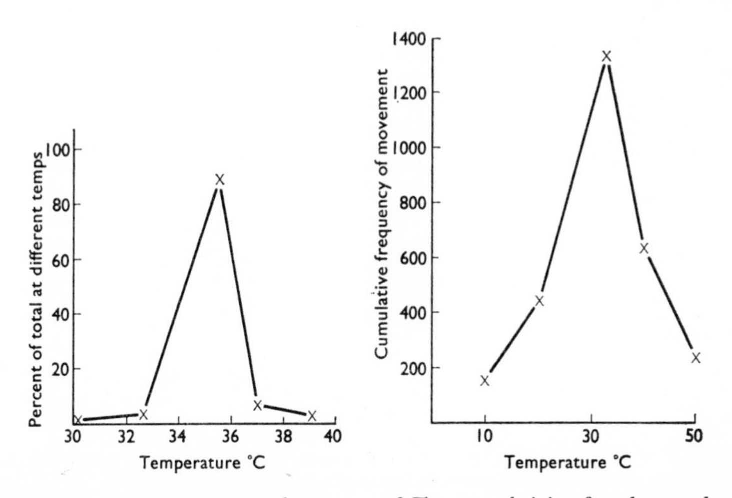

Figure 19 The eccritic thermal response of *Terranova decipiens* fourth stage larvae, and their activity with respect to temperature. (redrawn after Ronald, 1960).

In cod muscle the larvae are infective to seals, and they appear to survive best in the harp seal. Upon ingestion by a seal the thermal responses might cause the larvae to migrate from the tissues and so into the seal's intestine. The body temperature of most seals is about 38°C, but the body temperature of the harp seal is 35°C.

The thermal preferendum of the human louse is 30°C, the body temperature of the host about 36°C; the thermal preferendum of the pigeon louse however is 35°C, and the body temperature of the pigeon is 42°C (Rakshpal, 1959).

Cohn (1966) found that *Tylenchulus semipenetrans* survived best at 10°C in Israel, and the narcissus strain of *Ditylenchus dipsaci* was found to accumulate at 10°C when left in a thermal gradient of 2°C–30°C (Wallace, 1961). Croll (1967a) showed that acclimatization influenced the eccritic thermal response of *D. dipsaci*. When batches were stored for thirty days at constant temperatures of 10°C, 20°C and 30°C respectively, a statistically-significant proportion accumulated at the temperature in whch they had been previously stored (Figure 20). Acclimatized worms showed a peak activity at their eccritic temperatures. In both *T. decipiens* and *D. dipsaci* accumulation in a thermal gradient is at the same temperature as that of greatest activity. This is somewhat unexpected, as the temperature of peak activity may have been anticipated as that of greatest dispersal, unless the accumulation is the result of a klinokinesis.

It was outlined in the discussion on light responses that light could

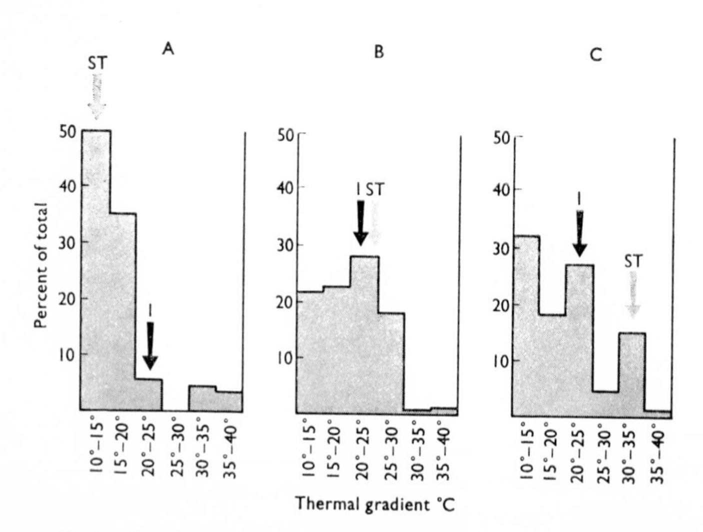

Figure 20 Acclimatization in the eccritic thermal response of *Ditylenchus dipsaci*. The distribution of *D. dipsaci* in a thermal gradient, following storage at constant temperatures for 30 days. A–10°C, B–20°C, C–30°C. I, is the inoculation point, S.T., the storage temperature. (Croll, 1967a).

act as a token stimulus, giving information regarding spatial orientation and day-night rhythms. Heat may have a similar importance as a token stimulus.

The classic examples of invertebrates which show directional movements in a thermal gradient are the ectoparasitic insects of warm-blooded hosts. Body lice (*Pediculus humanus corporis*) will move towards a heat source when placed in a thermal gradient. Similar positive thermotaxes have been described for a bed bug and the South American blood sucking bug *Rhodnius prolixus* (Wigglesworth & Gillett, 1934). *Cimex* and *Rhodnius* both perceive heat with their antennae, reacting not to radiant heat, but moving up a thermal intensity gradient. *Hirudo medicinalis*, unlike other leeches, moves towards a heated body (Herter, 1942) and is more comparable to nematodes as this leech is an ectoparasite in an aquatic medium.

There is ample evidence that some of the invertebrate parasites of warm-blooded hosts respond positively to a heat source, and the strong inference is that this response enhances the likelihood of transmission (page 2). The movement of infective *Nippostrongylus brasiliensis* larvae (Parker & Haley, 1960), and the many reports of positive thermotaxes for hookworm and *Strongyloides* larvae, may be explained as responses to token stimuli, aiding them in the location of their warm-blooded hosts. Although many of these worms occur in hot tropical areas, the apparent biological advantage of thermotaxes to adult parasites in cold-blooded hosts is not as readily explained. Both *Turbatrix aceti* and *Panagrellus redivivus* are bacterial feeders, and show movement towards a heat source (Croll, unpublished). In these cases it may be possible that heat is a token stimulus, leading nematodes to regions of bacterial decomposition.

6 | RESPONSES TO CHEMICALS

Nematodes are aquatic, either living in large water masses like seas and lakes or in thin fluid films, e.g. between soil particles or in dung or tissues. In solution or as gases chemicals are important in nematode environments. Like light but unlike heat, gravity and touch, chemical stimuli can vary in intensity and in quality. As there are many different chemical compounds, chemical stimuli can, therefore, be very specific. Through chemical perception nematodes may obtain information about their environment and the whereabouts of other organisms, plants and animals, including other nematodes.

> 'Obviously mononchs hunt by the aid of some sense other than sight, since both they and their prey usually live in subterranean darkness. Picture these ferocious little mononchs engaged in a ruthless chase in the midst of stygian darkness. We may imagine them taking up the scent of the various small animals upon which they feed, among which almost anything they can lay mouth to, seems not to come amiss and pursuing them with relentless zeal that knows no limit but repletion' (Cobb, 1917).

Host finding by plant-parasitic nematodes

The role of chemicals in orientation to roots has been much studied. Steiner (1925) observed aggregations of *Meloidogyne* larvae around roots, especially around root tips. His observations stimulated a search for similar aggregations of different nematodes and in different conditions. Two questions, in one form or another, have been continually asked—firstly, can nematodes locate plant roots through orientation responses, and second, if they can, how do they do it. The second of these questions i.e. the mechanism of orientation, is dealt with later (page 84). Borner (1960) published a review on root exudates, to which the reader is referred for other literature in this subject. Jones (1960) has summarized some of these factors; see Figure 21.

Because of the relative immobility of plants, the secretions of their roots can lead to an accumulation of chemicals to form concentration gradients. (Strictly the term 'concentration gradient' should be used only where the unit rise and unit slope, that is, the sine of the angle of rise, is constant (Hemmings, 1966).)

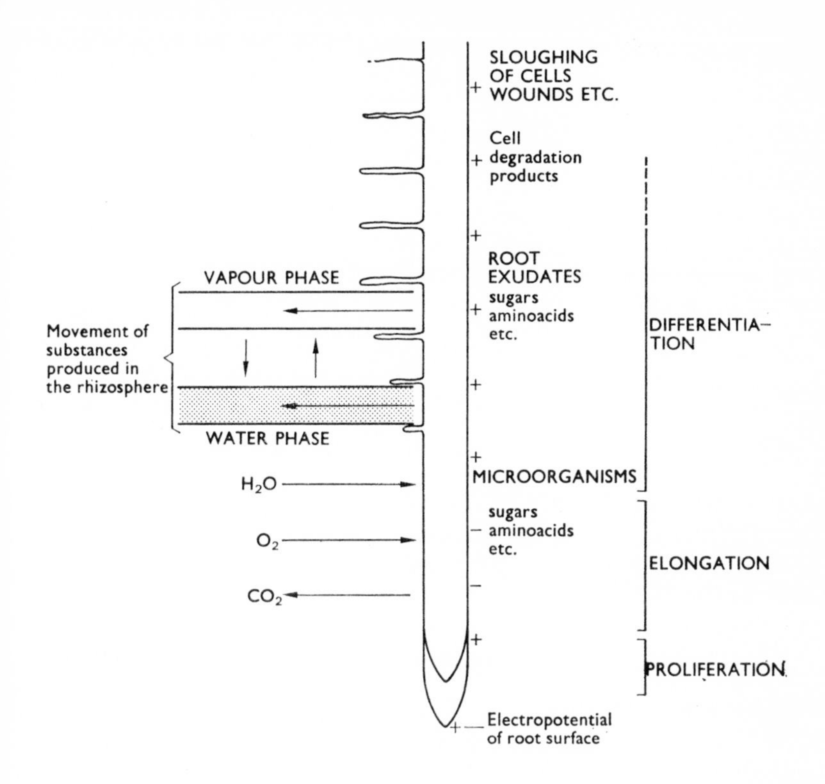

Figure 21 Diagrammatic representation of the root-soil complex, showing some of the factors that may affect nematode orientation to plant roots. (Jones, 1960).

Concentrations of carbon dioxide decrease, and concentrations of oxygen increase, with distance from the plant root, and so form gradients. Root exudates may also form gradients around roots.

Repeating Linford's (1939) experiments, Wieser (1955) found that *Meloidogyne hapla* was repelled by the apical 2 mm of tomato roots and attracted to the next 6 mm, while the following 8 mm of the root were neutral or slightly repellent. From experiments with egg plant and soybean, Wieser (1956) suggested the presence of attractive and repellent substances possibly associated with chemical decomposition of the root. Loewenberg *et al.* (1960) disagreed after finding a distinct attraction of *M. incognita* to the apical centimetre of root. Peacock (1961) also disagreed with Wieser, but confirmed that when root extension stopped, nematodes were no longer attracted.

Heterodera larvae appeared to orientate to a chemical secretion of the host plant (Wallace, 1958c), and larvae appeared to anticipate the formation of new rootlets, by collecting opposite the points of origin of endogenous lateral rootlets (Widdowson, *et al.*, 1958).

According to Baunacke (1922), *Heterodera schachtii* was attracted to substances given off by living plants and the secretions from seedlings were more effective than from fully grown plants. *H. schachtii* were activated and moved faster in contact with root diffusate (Wiescher, 1959). Luc (1961) also reported that *Hemicycliophora paradoxa* orientates to millet roots at distances of 20 to 40 cm.

H. schachtii larvae did not orientate to host plant roots in sterile conditions *in vitro* but in asterile conditions bacteria were present, some of which were attractive, others repellent (Bergman & van Duuren, 1959; 1959a). On this evidence it was suggested that chemicals secreted by bacteria in the rhizosphere surrounding the roots are the real attractive agents.

Besides the work in asterile conditions, Sandstedt and Schuster (1962) described how nematode larvae can become trapped in the aqueous films around roots, under plants on agar, a method frequently used. Despite alternative suggestions, reviewers of the subject tend to conclude that chemical attraction to roots does occur (Jones, 1960; Bird, 1962; Wallace, 1963; Klingler, 1965). Some nematologists support a 'random movement hypothesis', i.e. that host finding is essentially accidental, others support the 'root attraction hypothesis' (terms adopted from Bird, 1962).

Kuhn (1959) believed that orientation to host roots was a random process, and that upon contact with a root surface, nematodes remain on the root. His 'random movement hypothesis' was supported by Sandstedt, *et al.* (1961), Sandstedt and Schuster (1962) and Schuster and Sandstedt (1962). Rohde (1960) suggested that carbon dioxide concentrations greater than those of the atmosphere inhibited nematode movement, and that as larvae reached the immediate vicinity of plant roots they became inactive. The idea that aggregation occurred through inhibition supported the earlier beliefs of Kuhn (1959). Re-analysing Kuhn's data, Bird (1962) concluded that Kuhn's results (Table 4), provide evidence against the random movement hypothesis. Blake (1962) tracked *D. dipsaci* on agar as they approached oat seedlings (see Figure 22).

Of the chemicals around roots, carbon dioxide is quantitatively the most important (Klingler, 1965). *Meloidogyne hapla* and *M. javanica* larvae were not attracted to concentrations from 1 to 20 per cent carbon dioxide, buffered with bicarbonate at pH 6·7 to 8·0 (Bird, 1959). Bird bubbled carbon dioxide into molten agar, and at the temperatures used, carbon dioxide may have been much less soluble. Later Bird (1960) observed an accumulation of *M. javanica* and *Heterodera schachtii* larvae and adults, adult *Pratylenchus mingus*, pre-adult *Pratylenchus* sp. and various rhabditids, around a capillary source of gaseous carbon dioxide. By tracking *Ditylenchus dipsaci* in agar, Klingler (1959) found that this nematode moved

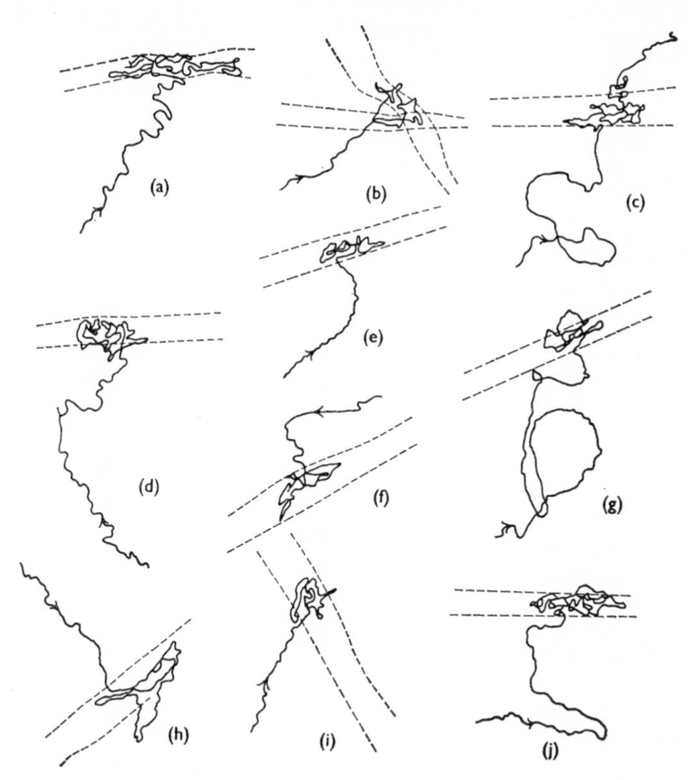

Figure 22 Tracks of *Ditylenchus dipsaci* in agar migrating across a sterile plate near an oat root, and remaining in the proximity of the root. The dotted lines show the position of the root. (redrawn after Blake, 1962).

towards a carbon dioxide source. In these experiments *D. dipsaci* entered the capillary tube from which the carbon dioxide was escaping!

Viglierchio (1961) placed tomato seedlings on the surface of experimental boxes, infested with *M. hapla* larvae with a dialysis membrane in between them. After 5 days a positive correlation was found between the position of the larvae and the position of the seedlings (Figure 23). In addition to the concentrations of larvae under the seedlings, there were lesser concentrations (Figure 23), centrally between the plants. Viglierchio suggested that attraction was small over 10 cm. From this and other experiments he concluded that a correlation of attraction with host preference or specificity was unwarranted (Viglierchio, 1961).

Ditylenchus dipsaci, *Meloidogyne hapla* and *M. javanica* and *Heterodera schachtii* all accumulated around sources of carbon dioxide, and so Johnson and Viglierchio (1961) suggested a positive chemical response (Figure 25). Increased carbon dioxide is associated with damaged roots,

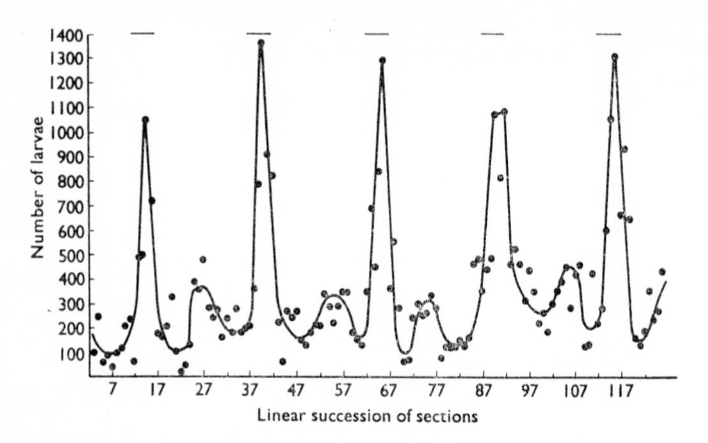

Figure 23 The distribution of *Meloidogyne hapla* larvae 5 days after being placed under tomato seedlings. The bars at the top of the graph indicate the plants. (Viglierchio, 1961).

and many have observed that nematodes attack injured parts of roots (Gadd & Loos, 1941). After being killed by heat, roots do not attract nematodes, but they become attractive again with the onset of decay (Bird, 1962). Most parasitic nematodes prefer healthy roots and move away from spent and injured tissues.

Lownsbery and Viglierchio (1961) obtained evidence that *M. hapla* accumulated around seed roots in response to a dialysable agent, thought to be carbon dioxide (Figure 23). These facts, together with the knowledge that the attractive agent has a low molecular weight (Viglierchio, 1961; Lownsbery & Viglierchio, 1961), is relatively non-polar and probably poorly dissociated, and that the attraction can be non-specific, led Klingler (1965) to emphasize the importance of carbon dioxide as an attractive agent for host finding in phyto-parasitic nematodes. Carbon dioxide is widely distributed in soil, and yet nematodes are able to attack only plant roots, additional compounds possibly more specific, would provide the integrated attraction behaviour.

Wireworms and the larvae of the weevil *Otiorhynchus* are also attracted to carbon dioxide (Klingler, 1965). Recently it has been found that alfalfa roots infected with *Fusarium oxysporum* produce more carbon dioxide than uninfected roots which may account for the increased attractiveness of *Pratylenchus penetrans* to infected roots (Edmunds & Mai, 1967). *P. penetrans* aggregated at the source of carbon dioxide (Edmunds & Mai, 1967).

Pitcher (1967) using time-lapse cinematographic techniques in an underground laboratory, observed and measured the accumulation of *Trichodorus viruliferus* around the tips of growing roots (Figure 24). Pitcher provides a review of the aggregation behaviour of seven species of *Tricho-*

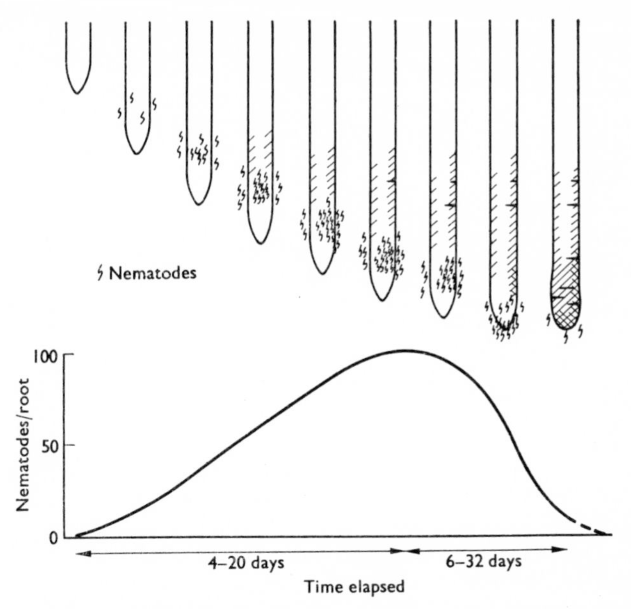

Figure 24 Schematic diagram of the build up of *Trichodorus viruliferus* on extending apple roots, showing nematodes massing around the elongation zone. (Pitcher, 1967).

dorus, and suggests that the root attractants must act at considerable distances.

Reducing agents such as sodium dithionite, cysteine, glutathione, and ascorbic acid attract *M. hapla* and *M. javanica* (Bird, 1959). Bird also concluded that these nematodes were not attracted to oxygen but *H. schachtii* and *D. dipsaci* responded positively in the experiments of Johnson and Viglierchio (1961). Klingler reported that *D. dipsaci* was attracted to the oxidizing agent potassium permanganate, and tyrosine attracted *M. javanica*. *Meloidogyne* larvae also responded positively to ascorbic acid, giberellic acid and glutamic acid (Bird, 1959). *Tylenchorhynchus martini* were tested on agar plates for attraction to a selection of inorganic salts, attraction was observed to 0·1M aluminium chloride and 0·25M cadmium chloride (Ibrahim & Hollis, 1967).

There is evidence that some plant parasitic forms, chiefly *Heterodera*, *Meloidogyne*, *Ditylenchus dipsaci* and *Pratylenchus* orientate to their host plants. The response may be galvanotactic (page 68), or it may be to chemicals. It has been suggested that nematodes respond to specific root exudates; much more evidence however supports the hypothesis that plant-parasitic nematodes respond to such non-specific stimuli as carbon dioxide. Oxygen, oxidizing agents and reducing agents are also attractive. The consensus is that attraction is non-specific (Lownsbery & Viglierchio,

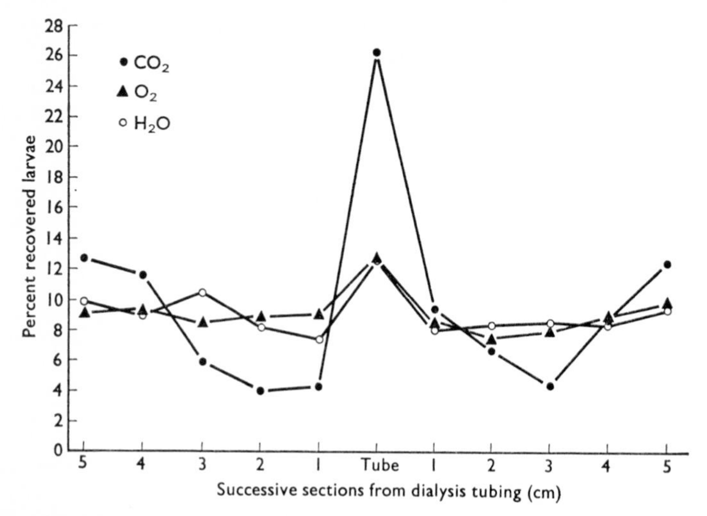

Figure 25 The distribution of *Meloidogyne hapla* larvae after a 24 hour exposure to concentration gradients of carbon dioxide and oxygen in dialysis tubing, with boiled distilled water control. (The carbon dioxide peak is significant over oxygen and water at the 1% level). (Johnson & Viglierchio, 1961).

1961; Wallace, 1963, and others), where accumulation around roots is probably the result of a balance between attractive and repellent substances. Bird (1962) suggested that future work should be done on a living host in a normal healthy state.

Chemo-sensitivity of microfilariae

Manson (1893) found that after feeding on an infected host, mosquitoes often contained more microfilariae than are present in an equivalent amount of blood obtained from a finger prick (O'Conner & Beatty, 1937). Harley (1932) speculated that the microfilariae respond chemotactically to the salivary secretion of the mosquito, injected before feeding. Strong, *et al.* (1934), found the same working with *Onchocerca volvulus* transmitted by *Simulium*.

A graphic description was published by Gordon and Lumsden (1939), of mosquitoes feeding on the North American leopard frog *Rana spheno cephala*. The microfilariae of *Folyella dolichoptera* were observed as they passed through the capillaries of the frog's webbed foot. They concluded that chemotaxes played no part in the entry of microfilariae of *F. dolichoptera* into the mosquito intermediate host.

Some animal ectoparasites feed directly from capillaries, others from veins or from pools of blood. Until more is known of the exact numbers and distribution of microfilariae in these microhabitats, the possiblity of chemotaxes in microfilariae remains an open question.

The detailed investigations of Hawking (for review see 1965), and others studying microfilarial periodicity, have shown that microfilariae respond to certain chemical compounds in the blood. Current thinking favours the idea that periodicity is based on the circadian rhythm of hosts, the microfilariae responding to rhythmic physiological changes in hosts. Hawking named the unknown factor that causes some microfilariae to remain in the capillaries of lungs, when not in the peripheral blood, the 'retention factor'.

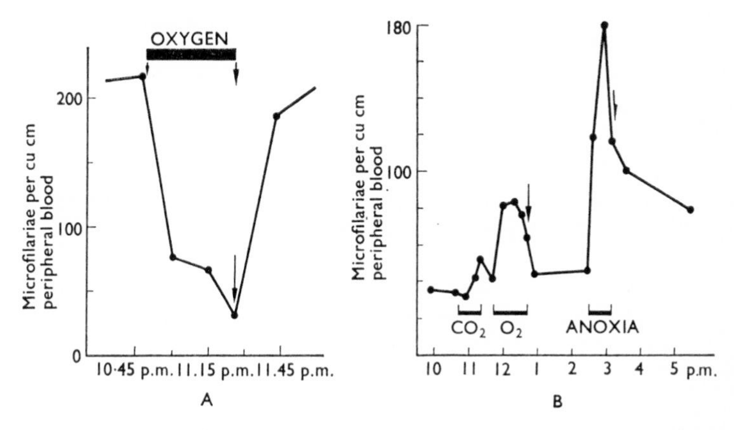

Figure 26 Some of the factors influencing the distribution of microfilariae in blood.

A. The effect of the host breathing oxygen on the number of *Wuchereria bancrofti* microfilariae in peripheral blood.

B. The effect of host anoxia on *Dirofilaria repens* microfilariae in peripheral blood of an unanaesthetised dog. Anoxia was produced by breathing a mixture of 12% oxygen plus 88% nitrogen. (redrawn after Hawking, 1962).

During the night the human circadian rhythm fluctuates, for example, changes in the oxygen tension of the blood and in the concentration of adrenalin. If a patient infected with *Wuchereria bancrofti* (causing elephantiasis), breathes oxygen at 230 mm mercury (normal oxygen pressure of air 160 mm mercury), the microfilariae rapidly disappear from the peripheral circulation, but return when the oxygen pressure of inhaled air is experimentally lowered. The concentration of microfilariae in the peripheral blood is also lowered if the carbon dioxide tension is lowered (Figure 26).

Adrenalin has little effect on the periodicity of *W. bancrofti*, but the adrenalin inhibitor Δ 9-a-fluorocortisol has striking effects. Acetylcholine is important in establishing and maintaining electrostatic potentials across membranes, and when injected into dogs infected with *Dirofilaria immitis*, causes a dramatic increase in the number of microfilariae in the blood. reaching a peak after 15 to 30 min and gradually returning to normal.

A number of compounds influence microfilarial periodicity, but no precise conclusions can yet be made. In different hosts, different species of microfilariae have developed responses that 'tune in' to different compounds or groups of compounds that fluctuate with the hosts' circadian rhythm. The older theories that the microfilariae are produced afresh each day, or are washed about passively in the blood with the dilations and constrictions of host capillaries, are largely discounted. Thus if one host is infected with both *W. bancrofti* and *Loa loa*, the microfilariae of which have opposed periodicities but are of very similar size, it seems unlikely that their rhythms can be controlled merely by passive movements of the vessels.

The evidence is that the microfilariae have a complex range of chemo-responses. If microfilariae of *Edesonfilaria malaysense* are transfused from one host to another at a different phase of the host's circadian rhythm, the periodicity of the microfilariae follows the cycle of the donor host for 8 h followed by 6 h adaptation after transfusion. Hence it seems that the microfilariae have their own endogenous rhythm (Hawking, *et al*., 1965). It may be concluded that chemicals affect microfilarial periodicity in such a way that the microfilariae must possess chemo-sensitivity. Nelson (1964) has reviewed the biology of filarial worms and the behaviour of the nematodes and arthropod hosts.

Chemo-sensitivity of free-living nematodes

Taniguchi (1933) examined the chemotactic responses of *Rhabditis filiformis*. Using a central well in an agar-filled dish, he tried various chemicals: 'At times the nemas intrude the diffused area and fall into the well while at other times they run away, having a distaste for the chemicals —the former and the latter negative.' Taniguchi reported an attraction to acids more than alkalis, and to inorganic acids more than organic. *Rhabditis filiformis* was attracted to alcohols but repelled by copper sulphate and silver nitrate. These rather empirical observations did not include measurements of concentration, possibly a significant factor. *Ditylenchus dipsaci* may be attracted to the amino acids glutamic acid and aspartic acid at concentrations of 1 in 100 000 and repelled at a concentration of 1 in 1000 parts.

Nematodes are thought not to orientate to a pH gradient (Bird, 1959; Bergmen & Duuren, 1959; Johnson & Viglierchio, 1961; and Klingler, 1961), so perhaps Taniguchi's observations on free-living rhabditids are worth repeating. Ellenby and Smith (1966) recently demonstrated a positive correlation between the pH and the population levels of *Panagrellus redivivus* cultures, in which the population increased and decreased as the acidity of the medium increased and decreased respectively (Figure 27). It is unknown whether this nematode, which may be affected by pH changes, responds directionally in pH gradients. Clapham (1931) thought *Dorylaimus saprophilus* responded negatively in sodium hydroxide, and 1 per cent sodium hydroxide was lethal.

After studying the influence of chemicals on the orientation of *Rhabditella* sp. larvae and *Toxocara canis* larvae, Gofman-Kadoshnikov, *et al.* (1955), concluded that chemoreceptors were better developed in rhabditids than in ascarids.

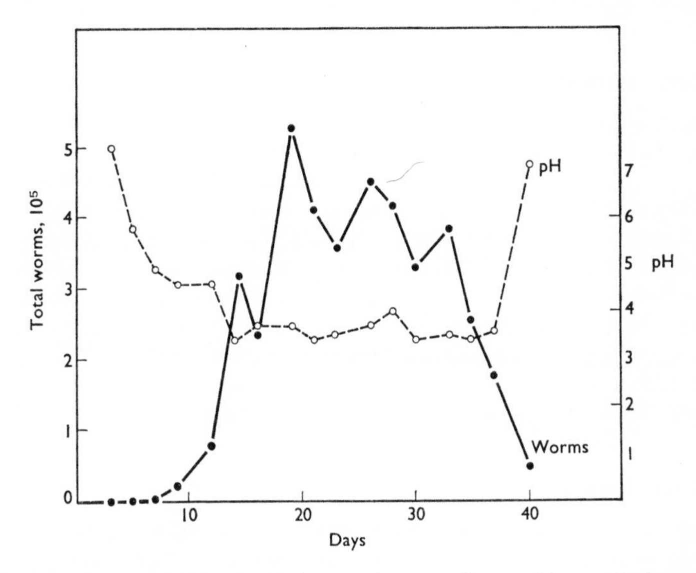

Figure 27 The variation in pH of the culture medium with population growth of *Panagrellus redivivus*. Acidity increases and decreases with population increase and decrease. (Ellenby Smith, 1966).

Chemo-sensitivity of marine nematodes

By culturing the marine fungi *Dendryphiella arenaria*, a deuteromycetous species and the ascomycete *Halosphaeria mediosetigera* in mycelial-cellulose mats, and leaving them on wooden stakes, Meyers and Hopper (1966) found that *Metoncholaimus* sp. was strongly attracted to the cultures. The cultures were staked out on sand flats which were covered even at low tide; after 1 to 3 days they were removed and examined. Large numbers of *Metoncholaimus* sp., most of which were gravid females, were recovered. One fungal mat submerged for 24 h contained ninety-eight gravid females, and one juvenile, while another submerged for 72 h contained 599 gravid females, eleven males and eleven juveniles. Smaller numbers of *Monhystera*, *Prochromadorella*, *Araeolaimus*, *Acanthonchus*, *Diplolaimella*, *Chromadora*, *Symplocostoma* and *Viscosia* were also present on the mats.

Metoncholaimus sp. swallows air bubbles, often rapidly; these accumulate in the intestine but gradually disappear as they pass backward, being completely absorbed from a half to two-thirds of the way along the intestine (Hopper & Meyers, 1966).

Oncholaimus vulgaris accumulated on pieces of torn mussel flesh, and Buerkel (1900) concluded that the nematodes had orientated by chemoreception at a distance.

Feeding stimuli

Workers attempting to culture nematodes *in vitro*, especially parasitic forms, have suggested the possibility of a 'feeding stimulus'. Hookworm larvae (Lane, 1930; Fulleborn, 1932), *Strongyloides stercoralis* larvae (Fulleborn, 1932), and *Nippostrongylus brasiliensis* (Cunningham, 1956; page 93) all become more active in host serum. Possibly a chemical constituent or constituents from serum are essential for activity and feeding. Serum has been found to provide a better feeding stimulus for *Ancylostoma caninum* than for *Nippostrongylus brasiliensis*. The identity of the active compounds is unknown and their existence is still a matter of conjecture. Nevertheless the possibility remains that they may be essential for successful *in vitro* cultivation (Roberts & Fairbairn, 1965).

Sex attractants

Sex attraction in nematodes is being studied increasingly. Greet (1964, also see Jones 1962), observed sexual attraction in *Panagrolaimus rigidus* (Figure 28). Copulation appeared to depend solely on chance encounters of males and females, but males never attempted copulation with other males. Using cellophane barriers, Greet (1964) separated males and females cultured separately, and examined their distribution 17 h later in two sides of a polyvinyl trough concluding (Figure 28) that each sex produced a secretion which attracted the opposite sex. He emphasized that the attractive substances serve only to bring the sexes together and that another set of substances or stimuli, some possibly tactile, evoke copulation.

Greet's (1964) results contrast with the observations on sex attraction in *Heterodera rostochiensis* and *H. schachtii* (Green, 1966) and *Pelodera teres* (Jones, 1967). In both genera only the males are attracted to the females. In *Heterodera*, Green concluded that both klinokinetic and klinotactic chemo-orientation occurred, in finding the females (Figure 29). The chemically attractive substance persisted on agar after the removal of the female and males were able to orientate to the chemical from at least

5 mm. Green was able to eliminate the possibility of other stimuli and he emphasized the significance of movement patterns in orientation (page 89).

Jones (1967) described a water soluble sex attractant, which activated and attracted male *Pelodera teres* to their females, but no attraction of the females to the males was observed.

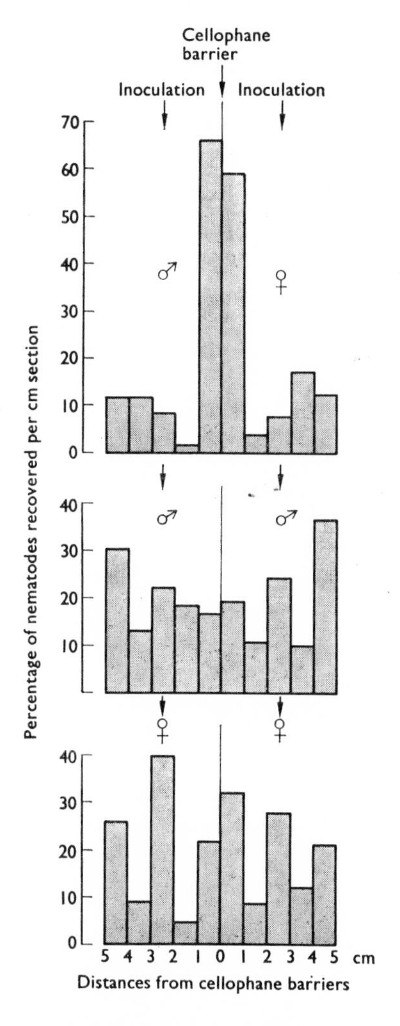

Figure 28 The distribution of *Panagrolaimus rigidus* in troughs after 17 hours at room temperature, demonstrating mutual sex attraction. The mid-line of each set of histograms represents the cellophane barrier between the two sections, arrows indicate points of inoculation. (Greet, 1964).

Adult male and female *Ancylostoma caninum*, introduced into the duodenal and ileo-caecal regions of the dog intestine, migrated to the mid-jejunum and copulated (Roche, 1966) but instead of sexual attraction each sex may have just been migrating to an optimal zone.

In studying sex attraction in adult *Trichinella spiralis*, Bonner and Etges (1967), demonstrated significant attraction *in vitro* in which males were more strongly attracted to females than vice versa. They suggested that thigmotaxes were important in attraction and copulation and adopted the term 'pheromone' from entomology, to describe the sex attractive substance. They also suggest that a sex repulsion may be present between like sexes. In none of the studies on sex attraction in nematodes has a chemical or group of chemicals been isolated.

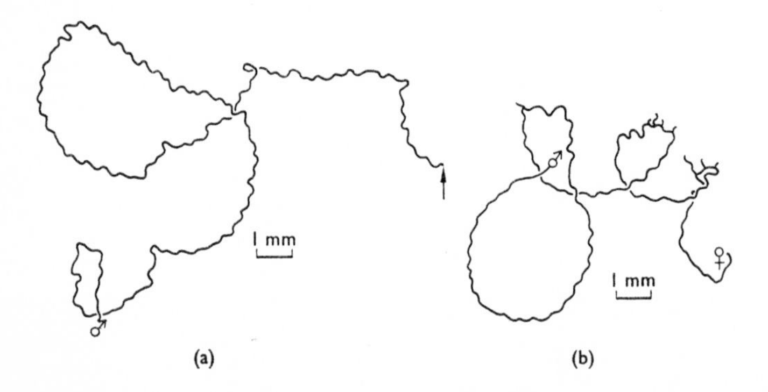

Figure 29 Sex attraction in *Heterodera*. a. The tracks left by an individual *Heterodera rostochiensis* male on agar with no female, the male moved from ♂ to the arrow. b. The track of a male *Heterodera schachtii* approaching a female. (Green, 1966).

Perception of chemical stimuli

A similarity has been observed between the sense organs of insects and nematodes (Crofton, 1966; Inglis, 1963), not one that reflects close phylogenetic relationships, but rather a similarity that has arisen through convergent evolution in the two groups. Both groups are covered by a cuticular exoskeleton, which has led to the adaptation of cuticular structures in some sensory organelles. In chemical perception for example it has been necessary in both the insects and the nematodes, that a nerve ending penetrate the cuticle, which is often found to be thinner than in adjacent areas.

For many years it was thought that cilia did not occur in nematodes but their presence has now been reported (Hope, 1964; Roggen, *et al.*, 1966; Ross, 1967). Roggen, *et al.*, found cilia in the amphids and in the somatic and cephalic papillae of *Xiphinema index*, the number of fibrils in the cilia however varied from the basic 9+2, for 9+2, 9+4, 8+2 and 8+4 were found. Cilia have also been described from the amphids of *Haemonchus contortus*, where they differ in different larval stages (Ross, 1967).

Amphids have been universally described as chemoreceptors, but electrophysiological evidence appears to be totally lacking for this assumption. Amphids are paired anteriorly situated organelles, lateral in position located immediately outside the ring of cephalid setae. Amphids are formed from cuticular invaginations which connect inwardly with an amphidial gland, which may itself run backwards parallel with the oesophagus. The amphidial nerve passes backwards, connecting the amphidial gland with the amphidial ganglia behind the nerve ring. Chitwood and Chitwood (1950) describe the sensory elements or specialized nerve endings of individual neurones, as 'terminals', the elements collectively being called a 'sensilla'. In most parasitic nematodes the amphids are greatly reduced and have lost the cuticular cavity.

Steiner (1955) described the 'cheeks' on the head of male root knot nematodes (*Meloidogyne*) as ampullae of the amphid. He believes that they increase the efficiency of chemoreception, which is especially well developed in the male—to locate the female. Experimenting with *Rhabditella* larvae, and *Toxocara canis* (page 61) Gofman-Kadoshnikov, *et al* (1955), concluded that amphids were chemoreceptors.

A positive reaction for esterases was observed in the amphidial pouches of all stages of *Meloidogyne javanica* and *M. hapla* (Bird 1966, plate 3). From this observation Bird postulated that these esterases may play an active part in the nematode's sensory perception, paralleling the acetyl choline action in the nervous system.

Marine nematodes possess large and complex amphids and would be excellent for experiments on chemosensitivity and Hope and Murphy (1969) have recently described some very unusual amphids. In the Enoploidea (Hyman, 1951) there are also 'cephalic slits', a pair of pouch-like structures, adjacent to the amphids which may be additional sense organs.

A model for the chemical basis of sensitivity

The sensitivity to carbon dioxide, shown by a number of plant parasitic nematodes has been outlined above. Carbon dioxide is also important in the exsheathment of some infective third stage larvae of animal parasitic forms. The infective third stage larvae of *Haemonchus contortus*, *Trichostrongylus colubriformis*, and *Nematospiroides dubius* undergo exsheathment (loss of the retained second stage larval cuticle) following entry into the host gut. The most important single compound which 'triggers' off exsheathment in these nematodes is carbon dioxide and/or undissociated carbonic acid (Rogers, 1960; 1962; Rogers & Sommerville, 1963;

Sommerville, 1964). Rogers (1966, 1966a) designed an elegant experiment which led him to make the following postulate.

He suggested that carbon dioxide reacts with two suitably placed sulphydryl groups on a receptor:

$$\text{—S—H} \quad \text{H—S—} + CO_2 \longrightarrow \text{—S—C—S—}$$

It was argued that iodine would block this reaction, by forming a disulphide link:

$$\text{—S—H} \quad \text{H—S—} + I_2 \longrightarrow \text{—S—I—S—}$$

In this oxidized state the disulphide link could again be reduced with hydrogen sulphide:

$$\text{—S—I—S—} + H_2S \longrightarrow \text{—S—H} \quad \text{H—S—}$$

which could return the receptor to its normal sensitive state.

Following treatment of the larvae with each of these compounds as outlined above, and subsequently feeding the larvae to their hosts, the postulate was confirmed. Although still a model, Rogers has apparently been successful in reversibly inhibiting the chemical receptor governing exsheathment and therefore the trigger for infection.

Croll and Viglierchio (1969) found that the positive response of *D. dipsaci* to carbon dioxide could be inhibited using an iodine treatment, and restored with subsequent hydrogen sulphide treatment. Unlike the 'trigger' mechanism of Rogers, the orientation response must require continuous sampling of the environment. The carbon dioxide concentration being determined from integrated pulses, according to a mass action equilibrium at the detector sites. Such a mechanism may be favoured by a spatial configuration detector for carbon dioxide. Thus iodine inhibition could be explained in terms of oxidation of a sulphydryl making the detector rigid and unable to conform to the carbon dioxide moiety.

Role of neurosecretion in chemosensitivity

The sensory cells of some of the labial and cephalic sense organs in *Ascaris lumbricoides* stain with paraldehyde-fuchsin and are thought to be neurosecretory (Davey, 1964), but the amphids were not equally fuchsinophilic. Davey (1964) describes dramatic changes in the quantity of neurosecretory material present following isolation from the host, which are thought to be correlated with massive sensory stimulation. The fact that the peripheral sense organs of *A. lumbricoides* are neurosecretory has led to the suggestion that there may be hormonal control in this nematode (Davey, 1964; 1966; 1967).

Soil, phytoparasitic, microfilariae and other stages of zoo-parasitic and marine nematodes respond to chemicals. Also the other evidence that host hormones affect parasitic nematodes implies that they possess chemoreceptors to perceive the hormones.

The nematicide, 2,4-Dichlorophenyl methanesulphonate, has an unusual mode of action, probably linked with an orientation response. The compound is highly specific for *Meloidogyne* species, but is apparently non-toxic. Experiments suggest that the nematicide makes treated roots repellent (McBeth, *et al.*, 1964).

7 | RESPONSES TO ELECTRIC CURRENTS AND MAGNETIC FIELDS

Over sixty years ago, Jennings reviewed the reactions of Protozoa to electric currents.

'It is interesting to note that this cramped and incoherent behaviour is found only under the influence of an agent that never acts on animals in their natural existence. The reaction to electricity is purely a product of the laboratory.' (Jennings, 1906.)

Fraenkel and Gunn (1940) were equally convinced that galvanotaxes or electrotaxes did not occur under normal field conditions. It was not known then that negative electrical potentials occur on root surfaces of many plants; and that these bioelectric potentials are of the same order as the physiological threshold for galvanotactic responses in nematodes. The root hairs of *Avena*, *Pisum* and *Zea* develop a negative potential of 80 to 115 mV, relative to a solution of potassium chloride (Etherington & Higginbotham, 1960).

The current and potential needed to evoke a response, or the galvanotactic threshold has been measured for a few species of nematode. *Turbatrix aceti*, *Heterodera schachtii*, and *Ditylenchus dipsaci* show directional movement to the poles at 20 to 30 mV (Jones, 1960). The larval stages of the animal parasite *Trichostrongylus retortaeformis*, when placed in a current of 10 to 40 mA, showed a 60 per cent movement towards the agar bridge cathode (Gupta, 1962). The threshold for orientation of *Panagrellus redivivus* was measured by Caveness and Panzer (1960) and found to be around 0·02 mA.

Table 4 Percentage of *Panagrellus redivivus* migrating to the cathode

Current (mA)	0	0.01	0.02	0.04	0.08	0.20	0.40	1.00	2.00
Per cent movement	54%	49%	73%	82%	69%	76%	86%	92%	86%

(Caveness & Panzer, 1960).

Bird (1959) suggested that the potential gradient produced in the vicinity of root hairs may assist plant nematodes to find them. Larval stages of *Meloidogyne javanica* were attracted to roots along a potential gradient produced by lowered redox potentials. Most were attracted to the greatest negative potential difference (105 mV). Bird (1959) also found that *M. hapla* and *M. javanica* were strongly attracted to reducing agents, especially sodium dithionite. Thus Bird's hypothesis has experimental support, although Jones (1960) suggested that electrical fields associated with such potentials are unlikely to extend to distances greater than a few millimeters from the root.

The potential developed around the growing point of roots is often higher than that over the rest of the root, and the electronegative properties of the cell sap may also influence galvanotactic orientation, and lead phytoparasitic nematodes to the growing points of roots. Jones (1960) stressed that the potential gradient, rather than the actual magnitude of the current, is important in galvanotactic orientation.

When Klingler (1961) used a platinum or copper wire electrode applied to an agar nematode suspension, the response of *Ditylenchus dipsaci* reported by Jones, 1960, did not occur. Small aggregations were however observed at the cathode with platinum electrodes, and copper anodes appeared to be more attractive to the nematodes.

In addition to nematodes listed above, the following genera move to the

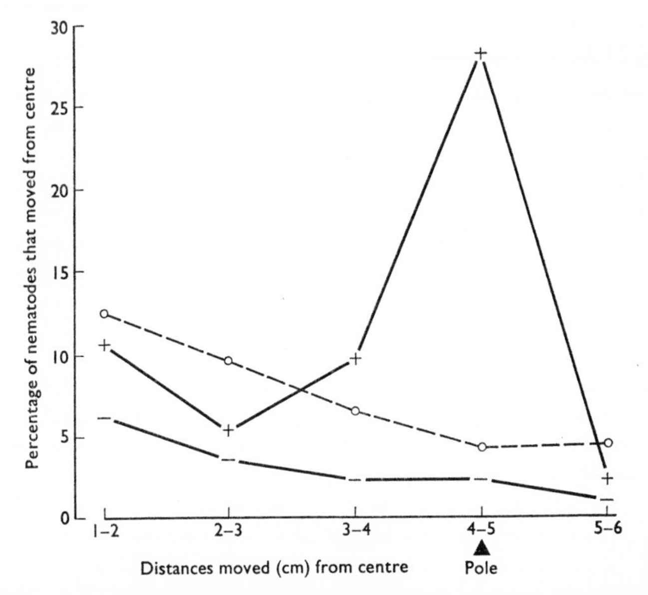

Figure 30 The response of larval *Heterodera schachtii*, to an electric current passed through a sand tap water system. The threshold being less than 20 mV/mm. +—+, potential applied from positive pole end; ——— potential applied negative pole end; 0—0 controls. (redrawn after Jones, 1960).

cathode when placed in an electric field: *Tylenchus*, *Rhabditis*, *Neocephalobus*, *Chiloplacus*, *Pratylenchus*, *Apelenchoides* and *Dorylaimus* (Caveness & Panzer, 1959).

Newly hatched larvae of *Heterodera schachtii* moved towards the anode when placed in an electric field (Figure 30) (Jones, 1960). When placed in a piece of flattened glass tubing with standard serum-tyrode, the orientation of *Litomosoides carinii* microfilariae was observed (Hawking, *et al.*, 1950). Potassium chloride-agar electrodes were used, and after 10 minutes in a current of 4 mA at 15 volts, all of the microfilariae were directed towards the anode. Reversal of the current completely reversed the orientation, and Hawking states that this orientation was quite independent of fluid movements. Orientations to the anode are rare in nematodes, and are somewhat unusual when compared with the galvanotactic reactions of protozoans, coelenterates, molluscs and chordates.

Using the larvae of the anisakid *Terranova decipiens*, from the muscles of cod, Ronald (1963) was unable to demonstrate a directional response at any voltage, pulse rate, pulse duration or amperage that he used, although he observed that the larvae became twisted and contorted in the current.

Whittaker (1969) found that larval *Pelodera strongyloides* migrated to the anode, at a threshold of 3 amp for dauer larvae and 32 amp for non-dauer larvae. Adult worms did not respond at currents of 1 amp to 10 amps.

Mechanism of galvanotactic orientation

Most workers have observed that nematodes squirm when the current is increased over the optimum level or the gradient suddenly reversed. Squirming has been interpreted as the direct influence of the current on the neuro-muscular system of the nematode, but is probably a re-orientation mediated through sense organs. The phenomenon of movement in an electric field is readily observed, but the mechanism of such a response is not well understood, and has been interpreted in a number of ways.

It is just possible that the galvanotactic phenomenon in very small forms is due to the charge on the nematode cuticle, resulting in an electrophoretic 'pull' on the nematode; this may explain why some nematodes appear to move towards the cathode while others move towards the anode.

The strong evidence against this in many reports is that only those nematodes actively moving, migrate to the poles, this suggesting that the electrophoretic pull is insufficient to cause migration. Once moving, the nematodes are more easily affected by a pull, but the size (20μ) of *Litomosoides carinii* microfilariae would seem to be small enough to permit electrophoretic drag of inactive microfilariae. Jones (1960), Hawking, *et al.*, (1950) and others have observed an immediate reversal in the direction of orientation in the nematodes, upon reversal of the poles, also suggesting a sensitivity rather than a purely electrophoretic phenomenon.

Many fish orientate to an electric field, and in these cases the low current and high mass of the organism make electrophoretic explanations virtually impossible!

If the galvanotactic response is a true orientation response it may be through special sense organs, or through the artificial electrical stimulation of some other receptor (perhaps amphids?). It may also be the result of a direct effect of the neuro-muscular system. As orientation itself is effected through differential contraction of muscle blocks, an artificial or external control of contraction could influence directional movements but would tend to be unco-ordinated.

Nematodes may first align themselves to the current and then move towards it as in the goldfish (*Carassius auratus*). Large goldfish also react more quickly than small ones (Nikolsky, 1963, page 60).

Jahn (1961) reviewed data collected over 27 years on the galvanotactic phenomena in protozoan ciliates, and his theory has received a wide acceptance (Dryl & Grebecki, 1965). Jahn argues that the protozoan cell acts as a core conductor in a volume conductor (the medium), and that the whole organism behaves like a nerve. Although this theory was developed for ciliates it might apply to galvanotaxes in nematodes. The nematode cuticle with its lipid layers could act as a good insulator, and the nematode tissues as a core conductor, conducting the current at a different rate from the medium. Whittaker (1969) speculated that the responses may be related to the metabolic rates of the different stages. This brief summary of the responses and the possible explanations leads to the proposal that nematodes are somehow 'aware' of currents and can orientate with respect to them.

The responses of nematodes to magnetic fields

Orientation to magnetic fields has been much studied in invertebrates (reviewed by Brown, 1965), and is well documented in insects (Picton, 1966) and molluscs (Barnwell & Brown, 1961; Charles, 1966). To date there are no reports of nematode responses to magnetic fields.

8 | RESPONSES TO GRAVITY

After a few days of culturing vertebrate faeces to obtain infective larvae of animal parasitic nematodes, the larvae swarm up the sides of the culture vessel (page 25). Large numbers of infective hookworm, strongyle and trichostrongyle larvae migrate from dung onto the surrounding grass, where they contact their host (Payne, 1922; 1923; Lucker, 1936; 1938; Stewart & Douglas, 1938; Rogers, 1940a; Rees, 1950).

The phytoparasitic nematode *Aphelenchoides ritzema-bosi*, parasitic on the aerial parts of plants, migrate from the soil. These and other examples provide evidence that nematodes move upwards in a vertical plane. Whether this is a real orientation response, through a sensory system or whether it is part of a random movement in all directions is uncertain.

Payne (1923a) concluded that the effect of gravity on hookworm larvae in a water film was negligible and Crofton (1954), emphasized that the surface tension forces acting on a small nematode are 10^4 to 10^5 times greater than gravity.

Vertical migration of animal parasitic larvae

Buckley (1940) measured responses of nematode larvae to gravity under controlled moisture, temperature and light. Using glass slides, he marked out sectors and spread a thin layer of acid washed sand over one side. After spraying on a little water, he placed the larvae in the middle of the slide, and stored it in a humid atmosphere, in a vertical position. After a suitable time he counted the larvae in each sector and calculated the amount of migration. Between 13°C and 14°C, but not at other temperatures, *Trichonema* larvae were strongly positively geotatic (Figure 31, also page 44). Although unexpected, he found that these migrations occurred when the slide was tilted with a slope of less than 1 in 8.

Crofton's results were directly opposite to those of Buckley. Crofton used a graticule eyepiece with concentric circles and, by moving a slide on his microscope through 90°, he was able to plot the positions of the larvae in a vertical plane. The circles were divided radially into four sectors and after placing larvae in the centre he counted the number in each sector after unit time (Figure 32). From this Crofton (1954) concluded that the movements of larval *Trichostrongylus retortaeformis*, mixed

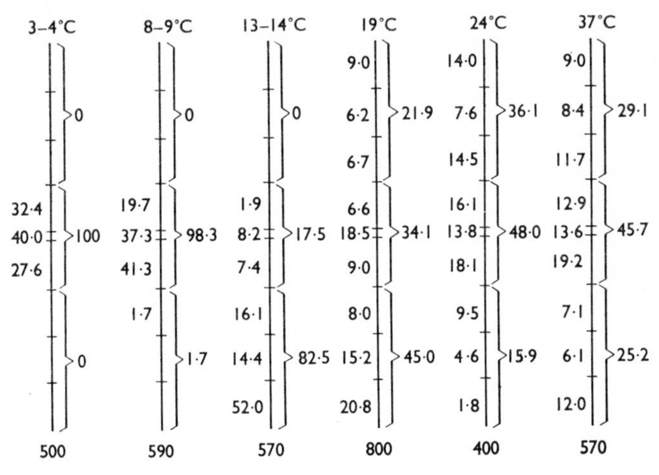

Figure 31 The vertical migration of infective larvae of *Trichonema* spp. at different temperatures after inoculation in the middle of the slide, after three hours in darkness. At the top is the temperature, and at the bottom the total number of nematodes. The figures at the sides show the percentage migration: left per centimetre, on the right per section. (Buckley, 1940).

trichostrongyle larvae from sheep, and *Trichonema* larvae from horses were uninfluenced by gravity. Broadbent and Kendall (1953) used Crofton's results (Figure 32) to make a model for 'random walk' in larval *T. retortaeformis*.

Finding that the vertical migration of nematodes was random, experiments were designed to explain the vertical movements of nematode larvae. The distance travelled in one direction, by larval *T. retortaeformis* in channels of varying width, was measured. The results (Table 5) led Crofton to conclude that the distance travelled by the larvae decreased as the width of the channel increased. By plotting the distance against the theoretical curve for a larva moving one unit per second in a channel too narrow to turn, and comparing this with results in different channels, it was found that in narrowing channels the distance travelled approached the theoretical curve (Figure 33). Larvae at the side of the channel frequently continued along it in a straight line. This may have been from a bias in sinusoidal movement such that the direction of transverse movement was slightly towards the side of contact (Gray, 1953).

The chance of a larva moving from the horizontal to the vertical plane would be directly proportional to the frequency at which the larva encounters vertical projections. Once on a projection the distance covered depends on the width of the path. A wide path would permit some random movement, and the distance travelled would be proportional to $\sqrt{t}$ where t is time (Figure 33), but in a narrow channel the distance covered would be proportional to t.

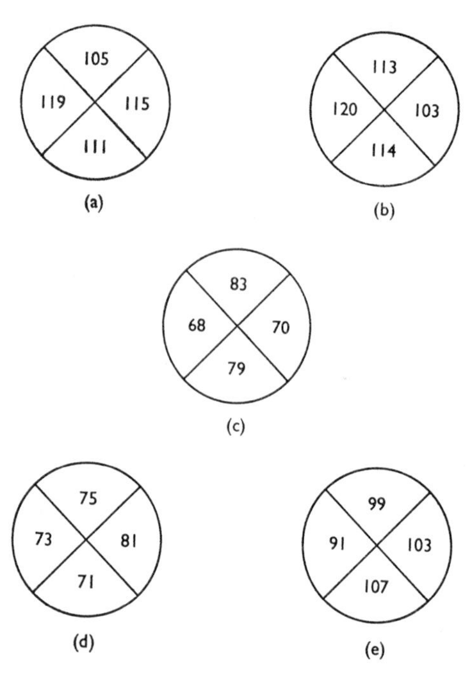

Figure 32 The distribution of larvae of some animal parasitic nematode larvae moving on a vertical plane, indicating no geotactic orientation. (a) *Trichostrongylus retortaeformis* at 17°C, (*b*) *T. retortaeformis* at 27°C, (c) mixed sheep trichostrongyles at 17°C, (d) mixed sheep trichostrongyles at 27°C, (e) horse trichonemes at 16°C. (data from Crofton, 1954; redrawn after Wallace, 1961b).

Crofton (1954) concluded therefore that where vertical migration of larvae was restricted as on grass and stems it may be described in terms of normal larval movements without reference to geotropism or the presence of undescribed special receptors.

Reesal (1951), concluded that the vertical migration of *Strongyloides agouti* was also the result of random movements, and *Dorylaimus saprophilus* was found to move irrespective of gravity (Clapham, 1931).

Vertical movements of *Mermis subnigrescens*

The gravid adult female of *Mermis subnigrescens* climbs grass stems on damp spring mornings to lay her eggs (page 38). Christie (1937) measured the vertical distribution of *M. subnigrescens* eggs on grass by counting the eggs, the number of pieces of vegetation at each horizontal level, and their height above ground, Table 6.

As may be seen from the table, most eggs were laid 10 in. above the soil, and the greatest number per available grass was at 10, 22, 20 and 14 in. respectively.

Lees (1953) concluded that the gravid females of *Panagrellus silusiae* climbed above the surface of the culture medium in response to a negative geotaxis. Ellenby and Smith (1966) studying the closely related *P. redivivus* partly contradicted these observations for, although the sides of the culture vessel contained mostly females, the ratio of males to females was in fact less than in the culture medium.

The stem and bulb nematode *Ditylenchus dipsaci* moves up and down in

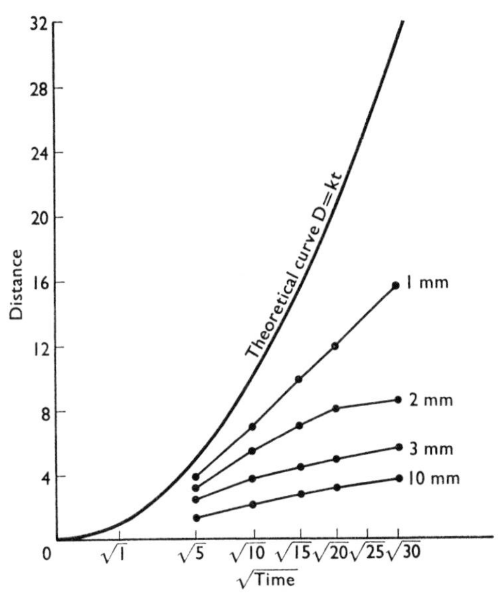

Figure 33 The mean distance travelled by larvae along channels of different widths. The theoretical curve D=kt represents the distance travelled by *Trichostrongylus retortaeformis* larvae, along a channel too narrow to permit a change in direction. (Crofton, 1954).

Table 5 Movements of larval *Trichostrongylus retortaeformis* along channels of varying widths (after Crofton, 1954).

Path width (mm)	Mean distance travelled in unit time				
	5 sec	10 sec	15 sec	20 sec	30 sec
1	3.9	7.1	9.9	11.8	13.5
2	3.3	5.4	7.0	7.8	8.3
3	2.6	3.7	4.3	4.7	5.2
10	1.4	2.1	2.7	3.1	3.3

Table 6 The vertical distribution of *Mermis subnigrescens* eggs following oviposition on grass (after Christie, 1937).

Height (in.)	Pieces available grass	Number of eggs	Eggs/grass
24	20	2	0.10
22	38	45	1.20
20	55	60	1.10
18	106	35	0.33
16	171	88	0.51
14	220	228	1.00
12	337	312	0.90
10	438	746	1.7
8	719	468	0.65
6	973	144	0.14
4	750	132	0.17
2	691	24	0.13

the soil with rainfall (Wallace, 1962; Figure 34). After rain most fourth stage larvae were at the surface of the soil; corresponding with the fact that *D. dipsaci* attacks plants at or above the soil surface. The absence of *D. dipsaci* at the soil surface during dry periods was probably not caused

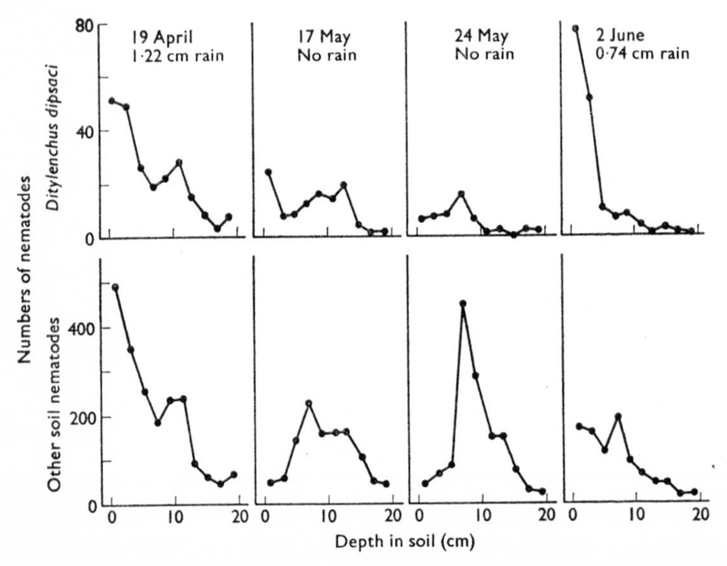

Figure 34 The vertical distribution of *Ditylenchus dipsaci* and other soil nematodes, in an infested oat plot, with respect to rain. The rainfall figured at the top of the graph was recorded in the week before sampling. (Wallace, 1962).

by a mortality from desiccation, as the fourth stage larvae withstand a relative humidity of 50 per cent, for 34 days (Wallace, 1962; Figure 34).

Although Wallace cites evidence for vertical movements, these could have resulted from responses to downwardly percolating solutions of such chemicals as oxygen, root emanations, or from downward movements associated with rainfall. The correlation drawn between rainfall and vertical migration does not necessarily imply a direct response to gravity (Wallace, 1962).

The most quoted example of a nematode geotaxis is probably the upward movement of the vinegar eelworm, *Turbatrix aceti* in a vinegar culture medium. Peters (1952) found that *T. aceti* was tail heavy and from this he postulated that upward movement was from the dragging effect on the tail.

Aphelenchoidesritzemabosi moved upward, and Wallace suggested it may have a negative geotaxis, a conclusion questioned by Barraclough and French (1965), when they observed that the orientation of *A. ritzema bosi* was random with respect to gravity.

Gravity is unique in that its direction and intensity are constant, and for this reason, only tactic movements can result from gravitational stimuli, geo-kineses being, by definition impossible. Some claim that nematodes orientate to gravity, but detailed studies like those of Crofton (1954), will probably confirm that the apparent responses to gravity are not via a sensory system. Nevertheless, the possibility of responses to gravity because statocysts or otoliths are unknown in nematodes, ignores the geotaxes described for digenetic miracidia and cercariae (Takahashi, *et al.*, 1961; Smyth, 1966).

9 | RESPONSES TO MECHANICAL STIMULI

Thigmotaxes are directional responses to mechanical stimuli such as touch, pressure and vibrations (Wallace, 1961b). Aggregation and swarming (page 25) of nematodes may be responses to tactile stimuli, which are more properly defined as thigmokineses. A positive thigmokinesis is a response which activates or inhibits movement through physical contact. It also includes that response which motivates the nematode to place the greatest possible area of its body surface against a solid substrate. The mechano-sensitivity may also be mediated through proprioreceptors in the muscles, or the cuticular outgrowths which reach their greatest development in marine nematodes (Figure 35).

The aggregation, swarming and related phenomena of nematodes are probably due, at least in part, to one or other of the following: thigmokinesis (Yoeli, 1957), cuticular adhesion (Ibrahim, 1967), movement (Gray & Lissmann, 1964) and chemical stimuli from the environment (Hesling, 1966).

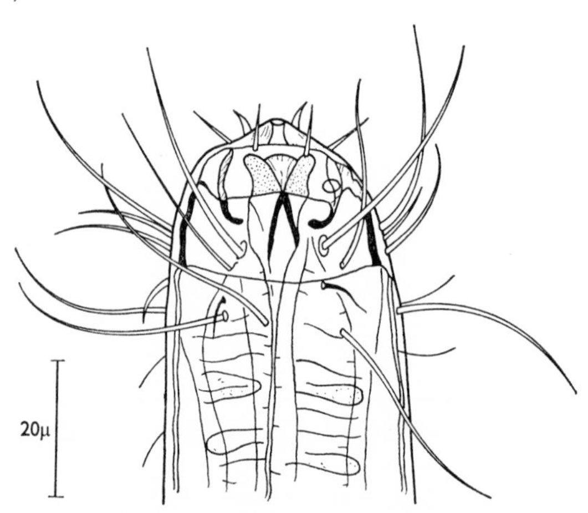

Figure 35 The anterior end of *Mesacanthoides sinuosus*, to show the elaborate setae typical of many marine nematodes. (redrawn after Wieser, 1959b).

Angle sense in nematodes

Another response related to touch stimulation in nematodes is an angle sense (Fulleborn, 1932; Lane, 1930; 1933), described in detail for hookworm larvae, which orientated at right angles to a solid flat surface with which they were presented. This response may aid the penetration of a host.

Penetration is also the biological role of second stage larvae of *Meloidogyne javanica*, and these larvae orientate at right angles to the host plant root (Wallace, 1968), as does *Hemicycliophora arenaria* (Van Gundy, 1961). *Heterodera* larvae also orientate to membranes (Dickinson, 1959).

Dorylaimus saprophilus actively avoided the point of a needle, and when touched in front they moved backwards and *vice versa* (Clapham, 1931).

Clapham's observation was further investigated by Croll and Smith (1970) using *Rhabditis* sp. They used a stimulus, measured in dynes/cm, and plotted the response with respect to the region of stimulation (Figure 36) and the intensity of the stimulus, and established adaptation. They interpreted the reaction as an escape mechanism from predators (Esser, 1963), the threshold of the stimulus approximating to that of contact with mites and predatory nematodes.

Thigmotaxes in aggregation

Thigmotaxes (or thigmokineses) have been described for the microfilariae of *Wuchereria bancrofti* in human blood (Yoeli, 1957). Blood from infected persons, examined 2, 4 and 8 h following venipuncture, contained

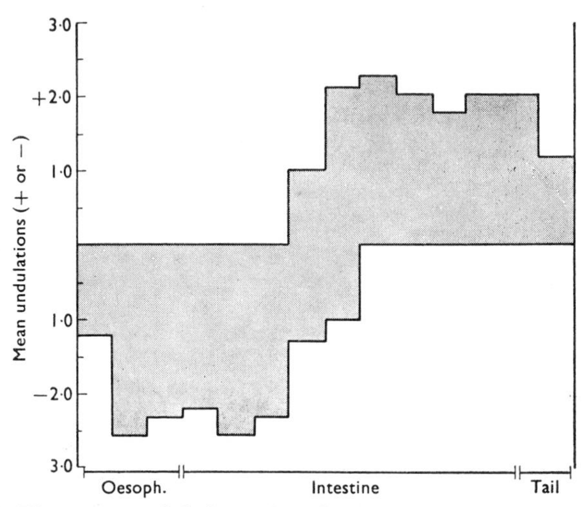

Figure 36 The mean undulations of *Rhabditis* sp. forwards (+) or backwards (−), when different areas of the body were given a mechanical stimulus of equal intensity. Each individual was immobile, and had not been previously stimulated. (Croll & Smith, 1970).

clumped masses of microfilariae termed 'medusa head formations' (Yoeli, 1957). The tails of the clumped microfilariae were directed towards the centre of the mass and each mass contained up to 150 microfilariae. In most clumps there was a central core, consisting of strands of fibrin or minute thrombocyte-leucocyte agglomerations, around which the nematodes were arranged. Clusters of these microfilariae were observed actively wriggling for up to 23 days in heparinated blood at 4°C.

When Franks and Stoll (1945) added the serum of an infected dog, the microfilariae of *Dirofilaria immitis* 'gradually slowed their wriggling and arranged themselves in orderly rows like matches in a box'. After further experiments these authors concluded that the phenomenon may have been due to surface tension.

Adult *Litomosoides carinii* in the pleural cavity of the cotton rat (*Sigmodon hispidus hispidus*) are usually found bound tightly together in a knot. This habit is also shared by the Gordian or Horse hair worms, the Nematomorphida, so called because of the similarity to the classical Gordian knot. If a small contact stimulus leads to great activity, Fraenkel and Gunn (1940) define such a response as a weak thigmokinesis. Organisms with such thigmokinesis aggregate together if other objects are absent.

Tactile stimulation and penetration

Dickinson (1959) examined the penetration of freshly-hatched *Heterodera schachtii* larvae into artificial nitro-cellulose (hydrophobic) and nitro cellulose (hydrophilic) membranes. It was found that penetration was a response to physical contact and could only occur through hydrophobic surfaces.

Mechanical stimulation may also initiate penetration of insect-parasitic nematodes.

Hatching, feeding, moulting, penetration and copulation are some of the many biological activities of nematodes. These and tactile orientations involve mechanical sensitivity and co-ordinated responses.

Responses of nematodes to fluid velocity gradients

If an organism in a fluid velocity gradient aligns itself parallel to the direction of flow, and the head points consistently in one direction, it is said to be positively rheotactic (Wallace, 1963). Such a response cannot occur in freely swimming nematodes in a clear medium, for although they are carried by the water current they are unable to refer to stationary objects. Weischer (1959) described a negative rheotaxis for larvae of *Heterodera rostochiensis*, but Wallace (1963) thought it was an orientation response. No rheotactic orientation to a water current was observed in *Ditylenchus dipsaci* (Wallace, 1961a).

Fulleborn (1924, 1932) thought the infective larvae of the animal-parasitic nematodes, *Ancylostoma caninum* and *Strongyloides stercoralis* had positive rheotaxes but Lane (1930) disagreed.

A positive rheotaxis was claimed for the chrysanthemum nematode *Aphelenchoides ritzema-bosi*, suggesting that the nematode migrates upwards in the downward percolating water of a chrysanthemum stem (Voss, 1930). However Wallace (1959b) found that water running down the stem actually decreased the upward migration.

No rheotactic behaviour has been confirmed in experiments. Rheotaxes may be established in the future, but no conclusions can be made at present.

Perception of mechanical stimuli

Labial and cephalic bristles or setae are extensions of the cuticle and are probably tactile receptors (Figures 35 and 38); each contains a nerve passing through the base of the bristle to join a papillary nerve. Such bristles are typically in marine nematodes and are almost always very small in soil and parasitic forms. It may be of some phylogenetic significance that the cuticular setae are arranged radially, contrasting with ocelli, amphids and deirids which are arranged laterally. The radial symmetry may reflect a primitive sessile state of attachment to the substrate by caudal glands.

The somatic setae of *Deontostoma californicum* are most frequently lateral and in the dorso-lateral and ventro-lateral planes (Hope, 1964). Deirids (page 85) may be tangoreceptors and it is possible that papillae are also sensitive to touch.

Inglis (1963) described campaniform-like organs from the marine chromadoroidean nematodes: *Cyatholaimus*, *Longicyatholaimus* and *Chonio-*

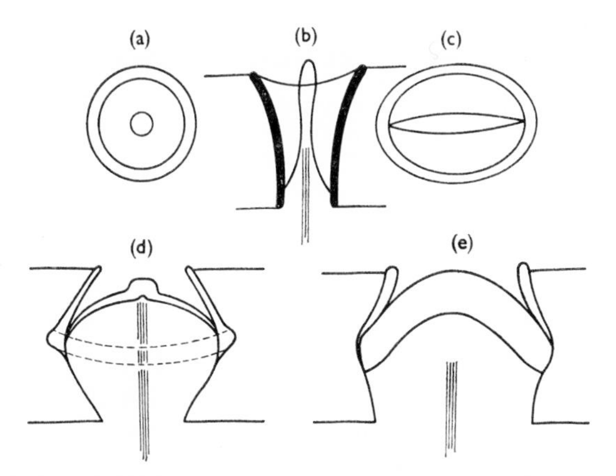

Figure 37 Various campaniform-type mechanoreceptors from marine nematodes of the family Cyatholaimidae. (Inglis, 1963). These structures have been reinterpreted by Wright and Hope (1968), see text.

laimus, all in the family Cyatholaimidae. There are two basic patterns of these mechanoreceptors, the commonest being an elliptical cup into which a thin sheet of cuticle projects, attached to the bottom and sides (Figure 37). These elliptical cups occur in longitudinal rows in all the genera listed above, and run the full length of the body. The receptors are associated with a nerve and gland cell, a common condition in nematode sense organs. The second form of mechanoreceptor, described by Inglis (1963), was from the marine nematode *Choniolaimus*, in which they are restricted to a short row of seven organelles running backwards from the amphids (Figure 37). These structures are considered to be mechanoreceptors, not on experimental evidence, but because of their structural similarity to the campaniform organs in insects (the similarity being one of convergence and not phylogenetic affinity). These observations have now been seriously questioned by Wright and Hope (1968) who studied *Acanthonchus duplicatus* using electron microscope techniques. They showed the campaniform-like organs to be ducts leading to pores through the cuticle.

From detailed microscopic studies, Maggenti (1964) concluded that the labial setae and/or papillae of *Deontostoma californicum* are sense organs associated with papillary nerves. The cephalic, somatic and caudal setae are sensory receptors connected with somatic nerves. Examining the lateral setae in detail, Maggenti differentiated between the seta, the scolopoid body and the bipolar sensory neuron. Scolopoid body (a term adopted from entomology) is synonymous with sensory plate, apical body or receptaculum. From the scolopoid body a filament extends through the cuticle and into the seta (Figure 38). The function of the scolopoid body is unknown, but Maggenti (1964) suggested it might be a reduced neuron

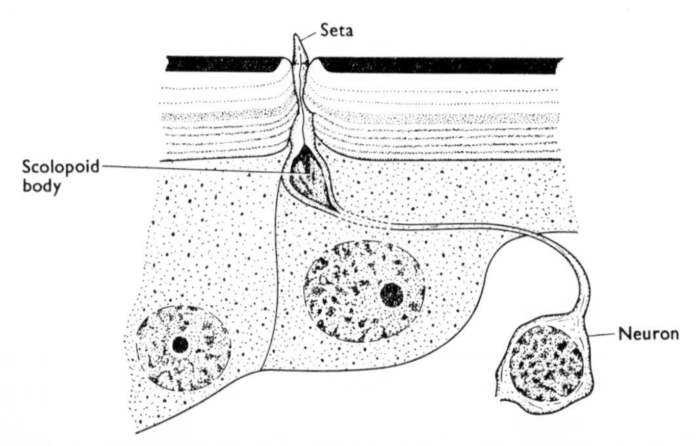

Figure 38 A sublateral hypodermal seta of the marine nematode *Deontostoma californicum* seen in longitudinal section, showing the scolopoid body and its connection to the neuron. (Maggenti, 1964).

that could amplify the external stimulus. The setae appear to connect with the lateral ganglia through the lateral nerves, and from observations on living decapitated worms, Maggenti postulated that an immediate reflex arc was present.

The somatic and cephalic setae of *Deontostoma californicum* connect to a complete, anastomozing peripheral nervous system (Croll & Maggenti, 1968). The density of neural fibres was correlated with the distribution of setae; plexuses were found in the cephalic, cervical and vulval areas, all regions where there are concentrations of setae (Figure 39).

Setae and bristles vary in form, but their functional significance has not been studied experimentally. The longer bristles on the head of the marine nematode, *Bathylaimus tarsioides* are jointed (Wieser, 1959), as are many others but their function is unknown.

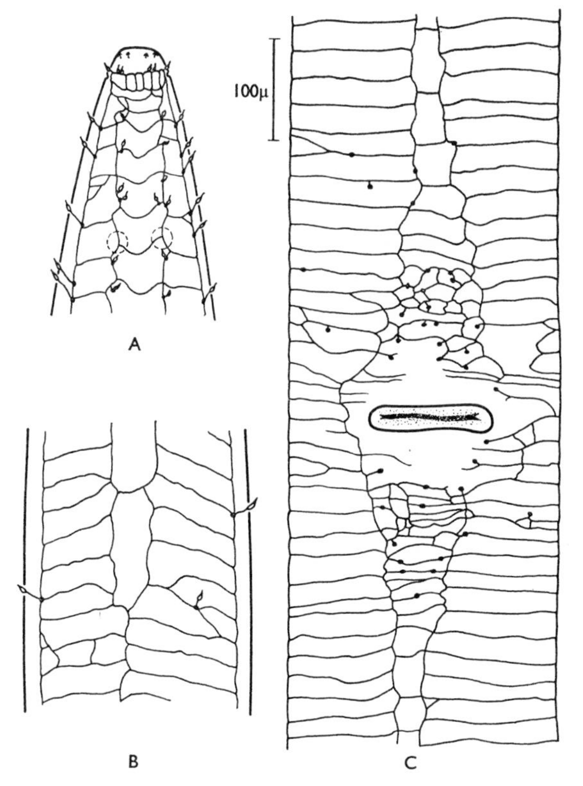

Figure 39 The peripheral nervous system of *Deontostoma californicum*, showing the neurofibrils and their connections with the somatic setae. A. Ventral view of the cephalic and cervical plexuses. B. Dorsal view showing the dichotomy of the dorsal peripheral nerve. C. Ventral view of the vulval plexus. (Croll & Maggenti, 1968).

10 | MECHANISM OF NEMATODE ORIENTATION

That nematode distribution is not random in chemical concentration gradients, or under the influence of certain undirectional stimuli, is no longer contested. The mechanisms by which orientation occurs, however, are still not properly understood.

There are two main opinions on orientation mechanisms although supporters of each disagree on details. Some believe that orientation is through directional responses or taxes, whereas others believe nematodes accumulate as the result of kinetic responses.

Orientation through kineses

Near suitable molluscan hosts, the behaviour of digenetic miracidia alters (Faust, 1924; and others). Davenport, *et al.* (1961), tracked the miracidia of *Schistosoma mansoni* in water and then in a water extract of the snail host *Australorbis glabratus*. From their results and those of others, has grown the notion that finding the mollusc is through a *klinokinesis*. The rate of turning increases in proportion to an increase in the concentration of snail mucus, or some other snail secretion.

Blake (1962) concluded that *Ditylenchus dipsaci* found oat seedlings by chemo-klinokinesis (Figure 22). The rate of turning of the nematode became greater as it entered an area of concentrated root exudate, or associated compound, so that it remained there until it contacted the host plant.

The implications of such a response are interesting, as the possession of a klinokinesis implies a 'memory', however simple, for klinokinetic orientation requires a comparison of stimulation intensities with respect to time. Evidence for a memory in nematodes is lacking, but as a primitive type of memory and learning has been repeatedly demonstrated in Protozoa, it is possible that it occurs in nematodes.

If a klinokinetic response is to result in orientation, sensory adaptation must occur (Fraenkel & Gunn, 1940), so that the distance travelled in one direction is greater than that travelled in the opposite direction. Sensory adaptation has been demonstrated in the light response of *Trichonema* spp. larvae (Croll, 1966a), but whether this is at sensory or synaptic level is unknown. Green (1966) also found that males of *Heterodera rostochiensis* and *H. schachtii* attracted to their females in high concentrations of the chemical sex attractant 'fatigued' and that they possessed a chemo-klinokinesis.

Trichonema spp. third stage larvae showed a photo-klinokinesis when crossing a light gradient (Croll, 1965), and elements of a klinokinesis occurred in the galvanotactic response of *Ditylenchus dipsaci* (Jones, 1960).

Apparent attraction may also be the result of a lowered or inhibited activity, that is, orientation through an *orthokinesis*. In their experiments Peacock (1959), Lownsbery and Viglierchio (1960) and Viglierchio (1961) arranged plants surrounded by a semipermeable membrane and grown in nematode-infested soil or sand. The nematodes were aggregated at the point of the membrane nearest the root. Klingler (1965) points out that such experiments show that the nematodes respond to a diffusable substance from a distance (page 52). If the nematodes are inactivated by increased concentrations of chemicals, they would collect around the membrane nearest the root. The response may not be the result of attractive substances but could be through inhibition of activity at the membrane. From experiments with *Heterodera rostochiensis* larvae, Kuhn (1959) concluded that orientation is the result of random wandering, and inhibition of movement when the root of the host is reached. He decided that plant nematodes were not attracted chemotactically to their hosts, and the host influence was exerted by changing the rate of larval movement. *Heterodera* sex attractants stimulate males to move faster.

Kuhn's (1959) conclusions are based on no actual measurements of change in the rate of movement of the larvae of *H. rostochiensis*. Weischer (1959) using the same species, reported that when the nematodes came into contact with the root diffusate they moved faster and farther.

Rohde (1960) suggested that carbon dioxide concentrations greater than those of the atmosphere inhibited nematode movement, and that as the worms reached the root surface they aggregated through a chemo-orthokinesis.

The problem of symmetry in nematode orientation

When the basic structure of a typical nematode with paired, anteriorly-situated sensory receptors is considered, it is apparent that the intensity of stimulation received by each receptor will differ only fractionally (Figure 40a). When, however, the sensory system in a moving nematode is considered (Figure 40b), the sectional area of the environment sampled is increased fifteen to twenty times.

It if is assumed that the side-to-side movement of undulatory propulsion is used to sample the environment, the extent of sideways movement (the amplitude of the wave) becomes significant. When the

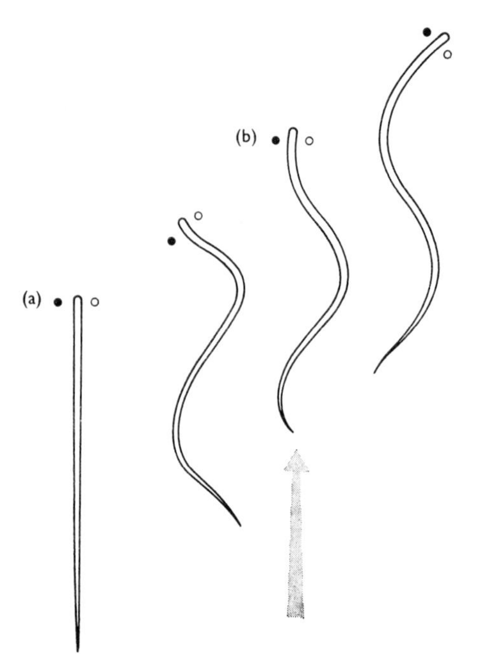

Figure 40 The arrangement of hypothetical sense organs, showing bilateral symmetry at rest (a), and during movement (b). When at rest the stimulus intensity at either side differs little, but when moving the amplitude of the wave-length increases the area of the environment sampled, and exposes alternate sides of the nematode to the stimulus. (Croll, 1967).

amplitude is great and the arc followed by the head is wide, the efficacy of sensory discrimination should be increased, while the rate of forward progression would be reduced. Therefore it may be deduced that a longer worm, or a worm with a greater wave amplitude, should have greater powers of sensory discrimination than a small nematode with a small amplitude.

The dorso-ventrally arranged longitudinal muscle blocks, used to propel nematodes, are thought to work in opposition to one another (page 8). For most nematodes this system causes nematodes to 'swim on their sides' on a flat surface; and even when moving in a 3-dimensional medium the parts of the body that are 'lateral' in the basic organization of the nematode, are arranged at right angles to the waves of the body. Amphids, ocelli, pigment spots, deirids and other organelles which may be sensory, are therefore not lateral but are situated one above the other during normal sinusoidal movement (Croll, 1967).

This greatly complicates the apparently simple sampling mechanism, for the plane of movement is at right angles to the plane of bilateral symmetry of the sense organs. This complication is most critical in a two-dimensional environment, when an undirectional stimulus is acting in the same plane. In a three-dimensional environment such as soil, dung, water, etc. sufficient variation in the plane may be introduced to allow the bilaterally symmetrical organelles to function. In water some nematodes corkscrew (Croll, 1969), and the model outlined above is further complicated by the greater versatility of movement at the head end, than in the general somatic musculature.

Orientation through taxes

When nematodes progress, they follow a characteristic sinusoidal path (page 8). Knowing that *Ditylenchus dipsaci* can only move if it waves its anterior end from side to side, it is difficult to understand Klingler's (1963) assertion on this evidence alone that *D. dipsaci* orientates to a germinating seedling through a klinotaxis, comparing the intensity of stimulation in its vicinity by alternate movements to the left and right.

Jones (1960) thought that the orientation of *D. dipsaci* to an electric potential may be klinotactic, as did Blake (1962) for the chemotaxis of *D. dipsaci*. In describing the sex-attraction of male *Heterodera rostochiensis* and *H. schachtii* to their females, Green (1966) also found elements of chemo-klinotaxes and stated that the receptors were too close to permit tropotactic orientation.

A nematode moving through any three-dimensional medium, such as soil, will be stimulated almost at random as it moves through the interstices between soil particles. Nevertheless, certain objections can be raised to an unquestioned acceptance that nematode taxes are klinotactic responses, an assumption based largely on the characteristic side-to-side movement.

Chromadorina viridis and *Panagrellus redivivus* can both swim freely in aqueous media, but *P. redivivus* is typically a crawler, and *C. viridis* more typically a swimmer. Their rates of activity vary considerably; *P. redivivus* moves at about 70 waves per min at 20°C, whereas *C. viridis* is more variable but can move at speeds of up to 500 waves per min at 20°C. Using the galvanotactic response of *P. redivivus* and the phototactic response of *C. viridis*, an experiment was devised to measure the efficiency of orientation when the stimulus was pulsated at different rates (Croll, 1967). If orientation were klinotactic, and the receptors sampled the environment at the extremes of sideways movement, it would be anticipated that effective orientation, and the degree of synchronization between stimulation and undulation, would be gradually lost as the interval between pulses increased. If, however, the mechanism of orientation were tropotactic, depending on continuous stimulation, effective orientation would be more likely to be suddenly lost at the point at which stimulation ceased to be physiologically continuous (Figures 41 and 42).

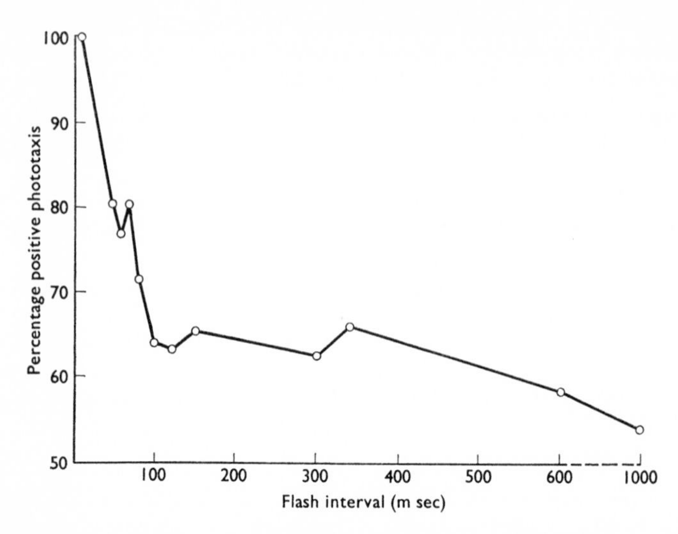

Figure 41 The efficiency of the positive phototaxis of *Chromadorina viridis*, in a flashing light of increasing interval between flashes, when the flashes have a constant duration of 10μsec. The light source was a xenon discharge tube of a transistorised stroboscope, emitting 450-460 mμ wavelength. (Croll, 1967a).

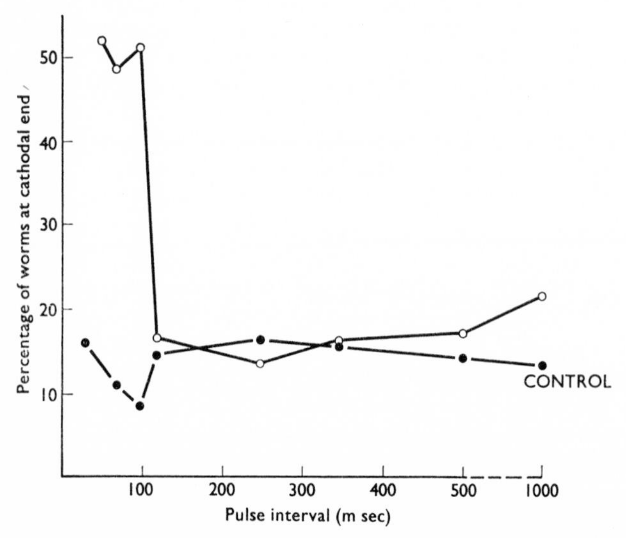

Figure 42 The efficiency of the galvanotactic response of *Panagrellus redivivus*, in a pulsating current, with increasing intervals between pulses. (Croll, 1967).

The sudden loss of effective orientation in the different nematodes, responding to different stimuli, was not correlated with wave frequency, but occurred at approximately 100 m sec pulse interval in both nematode species. This suggests the possibility of tropotactic orientation in these nematodes.

Movement patterns in nematodes

Most discussion on the mechanism of orientation has assumed that the environment is sampled as the nematode progresses towards the source of stimulation, often disregarding the importance of movement patterns. Tracks of nematodes in agar have been published by several authors, and in almost all, the tracks reflect a pronounced and persistent unilateral bias, the angle of lateral movement on one side being consistently greater than on the other side (Sandstedt, *et al.*, 1961; Rode & Starr, 1961; Klingler, 1961; Blake, 1962, and Green, 1966) A persistent unilateral bias was described for *P. redivivus* (Croll, 1969). The spiralling so typical of many nematode tracks, may be significant in nematode orientation as with the movement patterns of *P. redivivus* (Croll, 1969).

11 | GENERAL CONSIDERATIONS

In the previous pages some of the responses of nematodes to stimuli have been reviewed. Below an attempt will be made to integrate these observations. The responses of nematodes to individual stimuli, under laboratory conditions, are highly artificial, for in the field there is a complex interaction of changing environmental factors.

Categories of response of infective stages in finding their hosts

From responses described in the previous pages, a series of behavioural reactions may be recognized between the parasite and the host, which show an increasing degree of complexity. The total environment comprises: (1) the macroenvironment, (2) the microenvironment, or the host and its immediate surroundings (Figure 43). It is orientation responses to these aspects of the environment that guide nematodes to their hosts.

A HOST FINDING BY CHANCE

The theoretically primitive condition may be when food or host represent the same goal, and are found by purely random movement. This includes those movements of microbivorous soil nematodes in finding concentrations of bacteria and some mycophagous forms in their finding fungi. One specialization of the group could be the aggregation through orthokinesis around plant roots reported by Kuhn (1959). In these the phytoparasites move at random until they arrive at the host roots, where their activity is reduced or inhibited and they therefore accumulate.

B RESPONSES TO THE MACROENVIRONMENT

Many of the reported responses of nematodes are to widespread chemical and physical stimuli in the macroenvironment. These responses are not directly associated with host finding and are not responses to emanations of the host or its microenvironment. It is nevertheless through responses to these stimuli that infective nematodes enter the orbit of their hosts, indirectly enhancing the likelihood of parasitic transmission.

The negative geotaxis described for *Aphelenchoides ritzemi bosi* leads this phytoparasite to the buds and leaves of its plant hosts (Wallace, 1959).

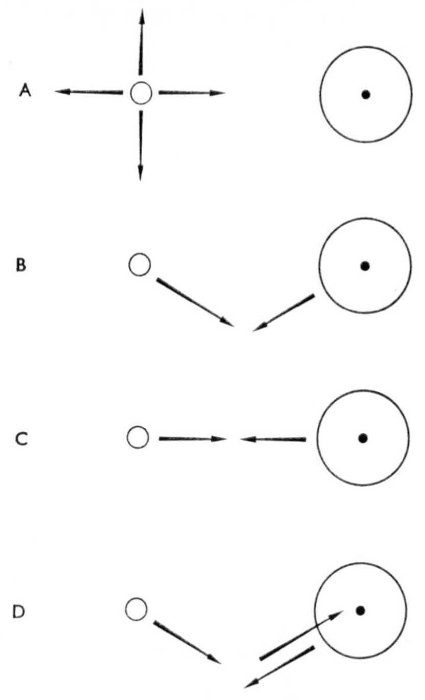

Figure 43 A model for the levels of behavioural interaction between nematode infective stages and their hosts. Left open circle = nematode, right dot = host, surrounded by the m icroenvironment of the host. For A, B, C, and D see text.

The photo-orthokineses and photo-klinokinesis observed in infective larvae of the horse strongyle, *Trichonema* spp. (Croll, 1965; 1966a) activate these nematode larvae and through influencing their behavioural pattern lead to vertical migration on the pasture. Most of these responses to macroenvironmental stimuli influencing the vertical migration of the infective larvae of ruminant parasites may be interpreted in this way (Rogers, 1940; Rose, 1950).

The gravid adult female of *Mermis subnigrescens* which migrates from the soil onto grass on damp mornings in early summer, is thought to be guided through a light-sensitive 'chromatrope' (Cobb, 1926, 1929; Christie, 1937; Croll, 1966b) which also controls the rate of oviposition (Cobb, 1929; Croll, 1966b). In the life cycle of this nematode, which is only parasitic in its larval stages, the female is the transmitting phase, laying her eggs on grass in the orbit of the grasshopper host.

C RESPONSES TO THE HOST MICROENVIRONMENT

The presence of a host exerts an influence on its surroundings; the area experiencing these influences is here defined as the host's microenvironment. The emanations may be physical or chemical and they may be secretions of widespread compounds (non-specific) or of compounds specifically associated only with a restricted host range (specific).

(1) *Non-specific stimuli*

Infective nematodes have been shown to respond to certain host-induced, non-selective, microenvironmental alterations. Actively penetrating larvae of warm-blooded hosts have been repeatedly shown to move up heat gradients: *Ancylostoma duodenale* and *Necator americanus* (Khalil, 1922; Payne, 1923; Lane, 1030; 1933), *Strongyloides stercoralis* (Fulleborn, 1924; 1932) and *Nippostrongyloides brasiliensis* (Parker & Haley, 1960).

The reports of galvanotactic responses in phytoparasitic nematodes (Bird, 1959; Caveness & Panzer, 1960; Jones, 1960) may be associated with host-finding through the bioelectric potential developed on host plant roots.

Carbon dioxide gradients increase and oxygen gradients decrease with proximity to plant roots. Johnson and Viglierchio (1961) demonstrated the accumulation of *Ditylenchus dipsaci*, *Meloidogyne hapla*, *M. javanica* and *Heterodera schachtii* around carbon dioxide and suggested a positive orientation response to it.

(2) *Specific stimuli*

The element of specificity in this subdivision limits the known responses to chemoresponses. Some hosts are believed to secrete specific chemical compounds to which nematodes have been found to orientate. Although many substances have been shown to increase hatching in *Heterodera rostochiensis*, it has been claimed that a specific hatching factor is produced by potatoes which stimulates larval hatching. Other exudates and secretions probably assist the larval nematode in locating the host root (Wallace, 1958; Rode, 1962). Specific stimuli may also activate larvae to hatch, and males to search for females.

It has been known since the time of Manson (1893) that after feeding on hosts infected with filarial nematodes, insect vectors often contain more microfilariae than in an equivalent volume of blood from a finger prick (O'Conner & Beatty). From these observations has grown the controversial suggestion that microfilariae may respond chemotactically to insect saliva.

D RESPONSES TO THE MACROENVIRONMENT, FOLLOWED BY SPECIFIC RECOGNITION RESPONSES IN THE MICROENVIRONMENT

In some infective nematodes at least two levels of response have been observed and this is probably true for many others. The first response is as in B, taking the nematode infective stage into the orbit of its host. Following contaminative entry of some trichostrongyles into their hosts, there is a specific recognition response to the microenvironment which stimulates exsheathment and development to the adult stage.

The importance of carbon dioxide to the exsheathment of the worms was shown by Rogers (1960) and Rogers and Sommerville (1963), and the carbon dioxide 'trigger' mechanism of the exsheathment stimulus by Rogers (1966, 1966a).

The chronological order of responses as in D (Figure 43) is similar to Wright's (1959) sequence of responses, which lead digenetic miracidia to find their snail hosts. As more information is gathered it is likely that the different phases in the life cycle will be shown to respond to stimuli most relevant to their biology (Cunningham, 1956; Croll, 1966e). The complex migrations shown by many animal parasitic nematodes in host tissues for example, are certainly influenced by stimuli within the host, and when understood, will need to be incorporated into the scheme.

The few responses reiterated above are merely representative of many which could be grouped under these headings. The examples demonstrate a more advanced exploitation of specific and non-specific responses to microenvironmental features in the phytoparasites than in zooparasites. In the past this has made the 'explanation' of certain responses of zooparasites difficult in terms of direct host location. In spite of their growth, the relative immobility of a plant host favours a build-up of chemical and physical gradients around itself, a condition less likely in mobile animal hosts. The formation of gradients around the plant host appears to be correlated with behaviour patterns directed to the host itself, through its microenvironment. In zoo-parasitic larvae, however, it is more common to respond to the macroenvironment in order to enter the orbit of the host.

Variation of responses at different stages in the life cycle

The relative dominance of individual responses may vary at different stages in the life cycle of any one species. Vertical migration is more usual in migrating larvae than in first stage larvae or parasitic adults, and sex attraction is limited to adults. Certain behavioural responses, therefore, are not so much characteristic of species but of stages. Cunningham (1956) examined a number of the responses of *Nippostrongylus brasiliensis*, at each of six stages in its life cycle; the first, second, and third larval stages, the early lung stage, preadults and adults. From his results he attempted to relate the behaviour of the stages to their biology (Figure 44).

N. brasiliensis adults responded most to thermal gradients (page 48 and Figure 44); preadults responded next and first stage larvae least. Cunningham (1956) also correlated the greater response of the larvae

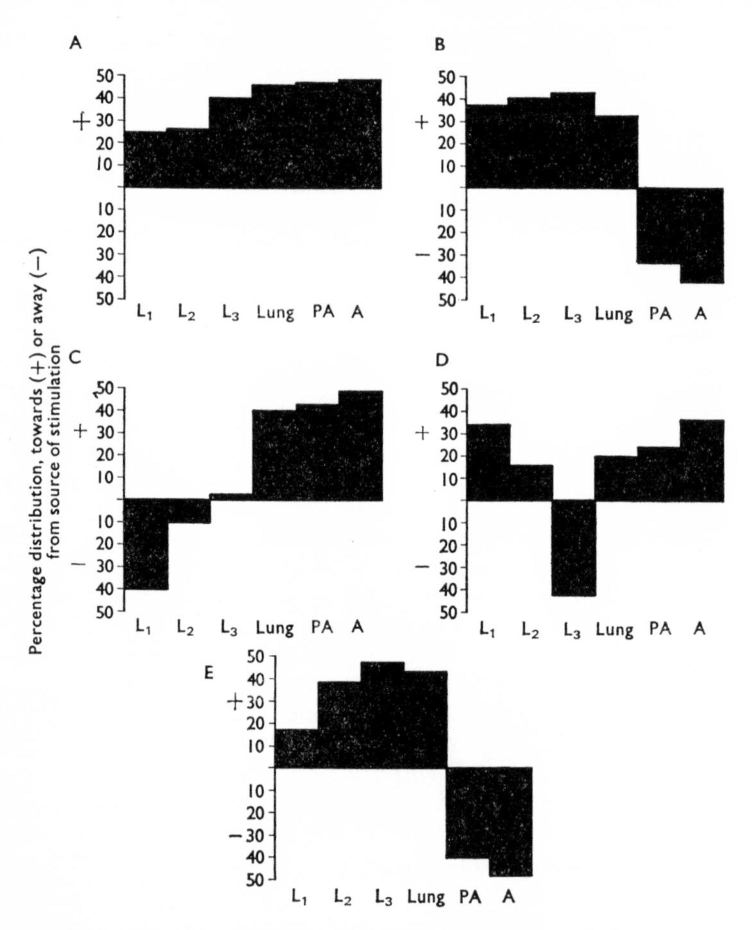

Figure 44 A comparison of the percentage distributions of various stages of *Nippostrongylus brasiliensis*, under directional stimulation. Even distribution is zero, so all of the larvae being on the side of stimulation is +50%. The stages used were: first, second and third stage larvae, lung stages, preadults and adults. A Heat, B Light, C Chemical (host serum), D Gravity, E Mechanical. (based on data of Cunningham, 1956).

than the adults to light with the different habitats of the two stages. His observations contrast with those of Africa (1931), who demonstrated a negative phototaxis in larval *N. brasiliensis* (Africa used sunlight, whereas Cunningham used electric light). The photic response of *N. brasiliensis* was also questioned by Parker and Haley (1960) who believed the apparent phototaxis was actually a thermotaxis to the heat from the light source.

The extent to which different stages of *N. brasiliensis* migrated in the vertical plane varied under constant conditions (Table 7), the infective larvae migrating the most.

Table 7 The negative geotaxes of different stages of *Nippostrongylus brasiliensis* (Cunningham, 1956).

Stage	Percentage moving with a negative geotaxis
Third stage larvae	71%
Second stage larvae	35%
First stage larvae	21%
Preadults	16%
Adults	2%

By comparing the responses to rat (host) serum, Cunningham concluded that more of the parasitic stages reacted positively to the serum than did the free living larvae.

Cunningham did not investigate the relative activities of the different stages, nor did he consider the efficiency of movement under his experimental conditions. Nevertheless, his results suggest that detailed study of the different stages in a single life history in other worms would be rewarding.

It is unknown whether combinations of stimuli act synergically or are mutually inhibiting, but knowledge of this kind would be useful in studying the physiological basis of sensitivity in nematodes.

Activity quanta in the behaviour of nematodes

Evidence is increasing that nematodes, like many other organisms, respond to environmental changes, which may be either increases or decreases in the intensity of stimulation. Wilson (1965) demonstrated the effect of a temperature change on *Nippostrongylus brasiliensis* larvae; and Bishop (1955, 1955a), recorded a greater rate of hatching of the eggs of *Heterodera rostochiensis* which were stored in fluctuating rather than in constant temperatures. *Trichonema* larvae remained inactive in constant illumination, but responded to changes (Croll, 1966a). These and other observations describe environmental changes which may release an energy 'quantum', which expresses itself as activity. The same or other stimuli may then influence subsequent orientation responses. The ease with which this quantum is released, or its magnitude once released, may vary much between nematodes. A suspension of plant parasitic nematodes almost invariably contains some that are active, and this proportion is remarkably constant. By comparison, a suspension of animal parasitic larvae from a ruminant, for example, may be quickly stimulated to activity and just as quickly become inactive. Although grossly over-simplified this simple observation may reveal a basic physiological difference in the mechanism and release of energy quanta.

'When free from disturbance the larvae of *Nippostrongylus muris* usually remain quiet inside the sheath, but the slightest stimulus from the ambient (temperature) awakens them to activity. They then extend their anterior ends into the air, the degree of extension being apparently directly proportional to the amount of stimulus applied'. (Africa, 1931).

Association between *Dictyocaulus viviparus* larvae and the fungus *Pilobolus kleinii*

The infective third stage larvae of *Dictyocaulus viviparus* use the widespread fungus *Pilobolus kleinii* for dispersal onto the pasture (Robinson, 1962; Robinson, *et al.*, 1962). At about the time that the *D. viviparus* larvae are infective, the mycelia of *P. kleinii* appear on the surface of the dung. The dispersal of the spores depends on the violent discharge of each sporangium, attached to the sporangiophore on the subsporangial swelling.

The fungus itself depends on ruminants, and the spores must pass through the gut of a herbivore before they can germinate. The sporangiophores are positively phototropic and, about midday, the black sporangia are shot towards the sun, following a trajectory which may carry them 10 feet. If the larvae make contact with the sporangiophore, they climb up onto the apical sporangium, where they become quiescent, and coiled. Robinson (1962) reports as many as fifty on a single sporangium.

Both *Cooperia punctata* and *Trichostrongylus colubriformis* have been found on the sporangia of *Pilobolus* spp. growing on laboratory egg cultures on dung. Up to thirty larvae were seen on a single sporanguim, and 444 larvae were projected in a single hour (Bizzell & Ciordia, 1965).

Other examples of behavioural sequences, involving a number of integrated responses, are sex attraction and mating, root finding and penetration in phytoparasitic nematodes, and the vertical migration of many animal parasitic nematodes.

Reversing movement

Reversing, in which the body waves are initiated at the tail and move forwards, appears to be common in nematodes. Green (1966) found that reversing was part of the orientation response of male *Heterodera rostochiensis* and *H. schachtii* to their females, although Croll (1969) interpreted this as a compensation movement used in re-aligning. *Panagrellus redivivus* reverses briefly at fairly constant intervals. The rate of forward progression in active worms is reduced if reversing increases until the reversing rate exceeds the forward progression, and tends to keep the nematode in the same place. If sensory stimulation of an inactive nematode releases an 'activity quantum' (page 95) which, once triggered off, leads to an inevitable period of activity, reversing movements could be important in reducing the rate of forward progression in unfavourable conditions. This may play a part in the retention of microfilariae in the lungs (Hawking, personal communication). Foragers like *Turbatrix aceti* and *Panagrellus redivivus* may require movement without migration, and reversing favours this (Carlson, 1962; Croll, 1969).

Factors in the control and mechanism of behavioural responses

Stimulation may be from external or internal stimuli (Figure 45); if from the former, external stimulation may act directly on the muscular and co-ordinating systems of the nematode, or indirectly through sensory receptors. The dermal light responses, certain temperature reactions and galvanotaxes, may all be the result of direct action, without sensory perception. When sensory perception is involved, acclimatization, habituation and interaction of stimuli may all influence the response.

Once stimulated, the physiological state of the nematode will determine the nature of any response, leading to the release of energy and the initiation of movement. Environmental factors, such as temperature,

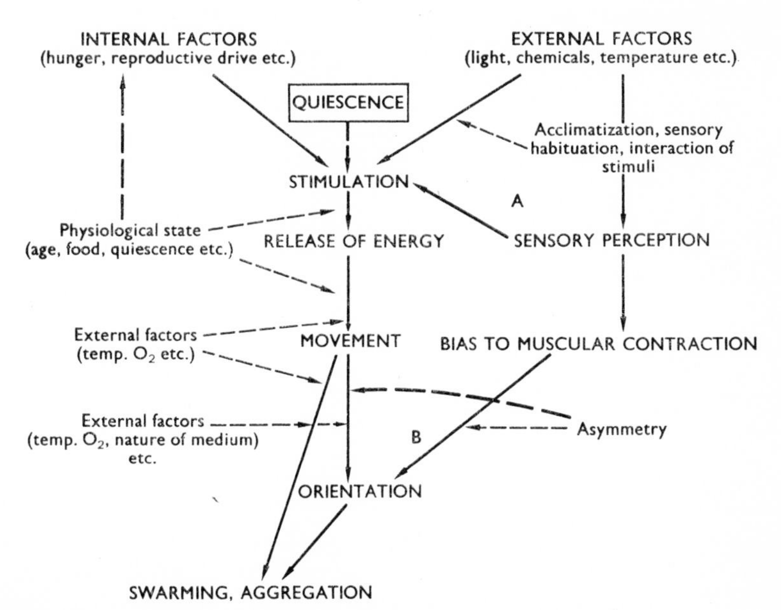

Figure 45 A model of the factors influencing nematode behaviour.

oxygen tension and viscosity of the medium, may all influence these processes.

Movement alone may be the result of these steps, i.e. 'thrashing', or it may lead to aggregation and swarming, through physical forces or sensory responses. It is at this point in behaviour that external stimuli can influence orientation through sensory perception. External conditions may either stimulate nematodes to become active (kinetic responses) or they may influence the pattern of movement and therefore the process of orientation.

Population density may affect many of these responses through: contact stimulation, sex attraction, accumulated excretory products, competition for food or oxygen, etc.

SELECTED BIBLIOGRAPHY

AFRICA, C. M. (1931). Studies on the activity of the infective larvae of the rat strongylid, *Nippostrongylus muris*. *J. Parasit.*, **17**, 196–206.

AKIMOTO, T. (1950) On the motion of *Ascaris lumbricoides* in a small tube. *Hirosaki Medical Journal*, **1**, 53–6.

ANDERSON, R. V. (1966). Observations on the nervous system of *Aporcelaimus amphidysis* n.sp. (Nematoda: Dorylaimoidea) *Can. J. Zool.*, **44**, 815–20.

ANONYMOUS (1964) Can you see with your fingers? *The Observer*, Feb. 2.

AOKI, T. (1959) Clinical observations on Dubini's ankylostomiasis and biological features of the mature larvae. *J. Nagoya Medic. Assoc.*, **80**, 1062–80.

AUGUSTINE, D. L., FIELD, M. E. and DRINKER, C. K. (1936) Observations on living *Microfilaria immitis* in the capillary circulation of bats. *Trans. R. Soc. Trop. Med. Hyg.*, **30**, 231–2.

BALASINGAM, E. (1964) Comparative studies on the effects of temperature on free-living stages of *Placoconus lotoris*, *Dochmoides stenocephala* and *Ancylostoma caninum*. *Can. J. Zool.*, **42**, 907–18.

BALDWIN, E. and MOYLE, V. (1947) An isolated nerve-muscle preparation from *Ascaris lumbricoides*. *J. Exp. Biol.*, **23**, 277–91.

BARNWELL, F. H. and BROWN, JR., F. A. (1961). Magnetic and photic responses in snails. *Experimentia*, **17**, 513–15.

BARRACLOUGH, R. M. and FRENCH, N. (1965) Observations on the orientation of *Aphelenchoides ritzemabosi* (Schwartz). *Nematologica*, **11**, 199–206.

BAUNACKE, W. (1922) Untersuchungen zur biologie und bekampfung der ruben nematoden *Heterodera schachtii*, Schmidt. *Arb. Biol.*, **2**, 185–8.

BERGMAN, B. H. H. and VAN DUUREN, A. J. (1959) Het bietencystenaaltje en zijn bestrijding. Vl De invloed van wortels van waardplanten en excreten hiervan op de bewegings richting van larven van *Heterodera schachtii*, in vitro. *Meded. Inst. Suikerbiet. Bergen-op-Zoom.*, **29**, 1–24.

(1959a) Het bietencystenaaltje en zijn bestrijding. Vll De werking van stofwisselings producten van sommige micro-organismen op de larven van *Heterodera schachtii*. *Ibid.*, **29**, 25–53.

BIRD, A. F. (1959) The attractiveness of roots to the plant parasitic nematodes *Meloidogyne javanica* and *M. hapla*. *Nematologica*, **4**, 322–35.

(1960) Additional notes on the attractiveness of roots to plant parasitic nematodes. *Nematologica*, **5**, 217.

(1962) Orientation of the larvae of *Meloidogyne javanica* relative to roots. *Nematologica*, **8**, 275–87.

(1966) Esterases in the genus *Meloidogyne*. *Nematologica*, **12**, 359–61.

(1967) Changes associated with parasitism in nematodes. Morphology and physiology of preparasitic and parasitic larvae of *Meloidogyne javanica*. *J. Parasit.*, **53**, 768–76.

BIRD, A. F. and WALLACE, H. R. (1965) The influence of temperature on *Meloidogyne hapla* and *M. javanica*. *Nematologica*, **11**, 581–9.

BISHOP, D. D. (1955) Hatching the contents of cysts of *Heterodera rostochiensis* with alternating temperature conditions. *Nature, Lond.*, **172**, 1108.

(1955a) The emergence of larvae of *Heterodera rostochiensis* under conditions of constant and alternating temperatures. *Ann. appl. Biol.*, **43**, 525–32.

BIZZELL, W. E. and CIORDIA, H. (1965) Dissemination of infective larvae of trichostrongylid parasites of ruminants from feces to pasture by the fungus *Pilobolus* sp. *J. Parasit.*, **51**, 184.

BLAKE, C. D. (1962) Some observations on the orientation of *Ditylenchus dipsaci* and invasion of oat seedlings. *Nematologica*, **8**, 177–92.

BLOOM, J. R. (1964) Photonegative reaction of the chrysanthemum foliar nematode. *Phytopathology*, **54**, 118–19.

BONNER, T. P. and ETGES, F. J. (1967) Chemically mediated sexual attraction in *Trichinella spiralis*. *Expl. Parasit.*, **21**, 53–60.

BORNER, H. (1960) Liberation of organic substances from higher plants and their role in·the soil sickness problem. *Bot. Rev.*, **26**, 393–424.

BOSHER, J. E. and McKEEN, W. E. (1954) Lyophilization and low temperature studies with the bulb and stem nematode, *Ditylenchus dipsaci* (Kuhn). *Proc. helminth Soc. Wash.*, **21**, 113–17.

BOURDILLON, A. (1950) Note sur le commensalisme des *Modiolaria* et des Ascides. *Bull. Lab. Arago.*, **1**, 198–9.

BRADLEY, C. (1961) The effect of certain chemicals on the response to electrical stimulation and the spontaneous rhythmical activity of larvae of *Phocanema decipiens*. *Can. J. Zool.*, **39**, 129–36.

BROADBENT, S. R. and KENDALL, D. G. (1953) The random walk of *Trichostrongylus retortaeformis*. *Biometrics*, **9**, 460–66.

BROWN, F. A. (1965) How animals respond to magnetism. *Discovery*, Nov.

(1965a) Circadian clocks. *Proc. Feldafing Summer School.* North-Holland Pub. Co., Amsterdam.

BROWN, K. N. (1954) Some observations on the behaviour of *Cercaria doricha* Rothschild (1935). *J. Helminth.*, **28**, 171–82.

BUCKLEY, J. J. C. (1940) Observations on the vertical migration of infective larvae of certain bursate nematodes. *J. Helminth.*, **18**, 173–82.

BUERKEL, E. (1900) Biologische studien uber die fauna der kieler Fohrde.

BURSELL, E. and EWER, D. W. (1950) On the reaction to humidity of *Peripatopsis moseleyi* (Wood-Mason). *J. exp. Biol.*, **26**, 335–53

CAMERON, T. W. M. (1923) On the biology of the infective larvae of *Monodontus trigonocephalus* (Rud.) of sheep. *J. Helminth.*, **1**, 205–14.

CAMPBELL, W. C. and TODD, A. C. (1955) Behaviour of the miracidium of *Fascioloides magna* (Basil 1875), Ward 1917, in the presence of the snail host. *Trans. Am. Micro. Soc.*, **74**, 342–7.

CARLSON, F. S. (1962) A Theory of Survival Value of Motility. Spermatozoan Motility, *A.A.A.S. Symposium* 72, Washington D.C.

CARTHY, J. B. (1958) *An Introduction to the Behaviour of Invertebrates.* Allen and Unwin, London.

(1964) Colour and Life. *Symposium Inst. Biol.*, **12**, 69–78.

(1966) *The Study of Behaviour.* Studies in Biology 3, Edward Arnold, London.

CAVENESS, F. E. and PANZER, J. (1960) Nemic galvanotaxes. *Proc. helminth. Soc. Wash.*, **27**, 73–4.

CAYROL, J. C. and RITTER, M. (1967). Importance et motivation des deplacements des Nematodes dans le sol et researches sur les migrations de *Ditylenchus myceliophagus*. *Revue de Zoologie Agricole et Applignee*, **66**, 92–102.

CHALAZONITIS, N. and CHAGNEUX, R. (1961) Photopotentials of the *Sepia* giant axon sensitized to light. *Bull. Inst. Oceanogr.*, 1223.

CHAPMAN, G. (1958) The hydrostatic skeleton in the invertebrates. *Biol. Rev.*, **33**, 338–71.

CHARLES, G. H. (1966) Sense organs (less cephalopods). *Physiology of Mollusca*, **2**, 455–521. Academic Press, London.

CHITWOOD, B. G. (1960) A preliminary contribution on the marine nemas (Adenophorea) of Northern California. *Trans. Am. microsc. Soc.*, **79**, 347–84.

CHITWOOD, B. G. and CHITWOOD, M. B. (1950) *An Introduction to Nematology*. Monumental Printing Co., Baltimore.

CHITWOOD, B. G. and MURPHY, D. G. (1964) Observations on two marine monohysterids, their classification, cultivation and behaviour. *Trans. Am. microsc. Soc.*, **83**, 311–29.

CHRISTIE, J. R. (1937) *Mermis subnigrescens*, a nematode parasite of grasshoppers. *J. agric. Res.*, **55**, 353–64.

CLAPHAM P. A. (1931) Observations on the tropisms of *Dorylaimus saprophilus* and *Rhabditis succaris*. *J. Helminth.*, **9**, 29–38.

COBB, N. A. (1914) Nematodes and their relations. *Yearbook Dept. Agric.*, 455–90.

(1917) The Mononchs (*Mononchus*, Bastian) a genus of free-living predatory nematodes. *Soil Sci.*, **3**, 431–86.

(1925) Organs of Nemas. *J. Parasit.*, **11**, 222–3.

(1926) The species of *Mermis*, a very remarkable group of nemas infesting insects. *J. Parasit.*, **13**, 66–72.

(1929) The chromatropism of *Mermis subnigrescens*, a nemic parasite of grasshoppers. *J. Wash. Acad. Sci.*, **19**, 159–66.

COHN, E. (1966) Observations on the survival of free-living stages of the citrus nematode. *Nematologica*, **12**, 321–7.

CONINCK, DE L. (1965) *Traite de Zoologie*. IV, Fascicule 2, Nematodes 586-731. Ed. Grasse P.P.

CROFTON, H. D. (1948) The vertical distribution of the infective larvae of *Trichostrongylus retortaeformis* in relation to their habitat. *Parasitology*, **39**, 17–25.

(1948a) The ecology of the immature phases of trichostrongyle nematodes. II. The effect of climatic factors on the availability of the infective larvae of *Trichostrongylus retortaeformis* to the host. *Parasitology*, **39**, 26–52.

(1954) The vertical migration of infective larvae of strongyloid nematodes. *J. Helminth.*, **28**, 35–52.

(1966) *Nematodes*. Hutchinson, London.

CROLL, N. A. (1965) The klinokinetic behaviour of infective *Trichonema* larvae in light. *Parasitology*, **55**, 579-82.

(1966) The light sense in nematodes. Ph.D. Thesis, London University.

(1966a) Activity and the orthokinetic response of *Trichonema* larvae in light. *Parasitology*, **56**, 307–12.

(1966b) A contribution to the light sensitivity of the *Mermis subnigrescens* chromatrope. *J. Helminth.*, **40**, 33–8.

(1966c) The chemical nature of the pigment spots of *Enoplus communis*. *Nature, Lond.*, **211**, 859.

(1966d) *The Ecology of Parasites*. Heinemann, London.

(1966e) The phototactic response and spectral sensitivity of *Chromadorina viridis* (Nematoda, Chromadorida), with a note on the nature of the paired pigment spots. *Nematologica*, **12**, 610–14.

(1967) The mechanism of orientation in nematodes. *Nematologica*, **13**, 17–22.

(1967a) Acclimatization in the eccritic thermal response of *Ditylenchus dipsaci*. *Nematologica*, **13**, 385–9.

(1969) Asymmetry in nematode movement patterns and its possible significance in orientation. *Nematologica*, **15**, 389–94.

CROLL, N. A. and MAGGENTI, A. R. (1968) A peripheral nervous system in Nematoda, with a discussion of its functional and phylogenetic significance. *Proc. helminth. Soc. Wash.*, **35**, 108–15.

CROLL, N. A. and VIGLIERCHIO, D. R. (1969) Reversible inhibition of chemosensitivity in a phytoparasitic nematode. *J. Parasit.*, **55**, 895–6.

CROLL, N. A. and SMITH, J. M. (1970) The sensitivity and responses of *Rhabditis* sp. to peripheral mechanical stimulation. *Proc. helminth. Soc. Wash.*, **37**, 1–5.

CUNNINGHAM, F. C. (1956) A comparative study of tropisms of *Nippostrongylus muris* (Nematoda: Trichostrongylidae). *Biol. Stud. Cath. Univ. Amer.*, **36**, 1–24.

DAVENPORT, D., WRIGHT, C. A. and CAUSLEY, D. (1962) Technique for the study of the behaviour of motile micro-organisms. *Science*, **135**, 1059–60.

DAVENPORT, H. E. (1945) Haemoglobins of *Ascaris lumbricoides* var. *suis*. *Nature, Lond.*, **155**, 516–7.

DAVEY, K. G. (1964) Neurosecretory cells in a nematode *Ascaris lumbricoides*. *Can. J. Zool.*, **42**, 731–4.

(1966) Neurosecretion and Molting in Some Parasitic Nematodes. *Am. Zoologist*, **6**, 243–9.

(1967) Endocrine basis for ecdysis in a parasitic nematode. *Nature*, **214**, 737–8.

DEAN, C. G. (1967) (personal communication).

DICKINSON, S. (1959) The behaviour of larvae of *Heterodera schachtii* on nitrocellulose membranes. *Nematologica*, **4**, 60–6.

DONCASTER, C. C. (1964) Four patterns of observation chamber for studying nematode behaviour. *Nematologica*, **10**, 60–6.

DONCASTER, C. C. and WEBSTER, J. M. (1968) Clumping of the plant parasitic nematode *Ditylenchus dipsaci* in water. *Nematologica*, **14**, 131–6.

DROPKIN, V. H. (1965) Behaviour of nematodes in relation to that of other invertebrates. *Nematologica*, **11**, 1–3.

DRYL, S. and GREBECKI, A. (1965) Recent Advances in research on excitability of ciliates. *2nd Intern. Cong. Protozoology*, London. Excerpta Medica Foundation, London.

EARL, P. R. (1959) Filariae from the dog *in vitro*. *Ann. N.Y. Acad. Sci.*, **77**, 163-75.

EDESON, J. F. B. (1960) The microfilariae of the periodic and semi-periodic forms of *Brugia malayi*. *Trans. R. Soc. Trop. Med. Hyg.*, **54**, 570–1.

EDMUNDS, J. E. and MAI, W. F. (1967) Effects of *Fusarium oxysporum* on movement of *Pratylenchus penetrans* towards alfalfa roots. *Phytopathology*, **57**, 468–71.

EDWARDIO DI, A. A. (1960) Time-lapse studies of movement, feeding and hatching of *Pratylenchus penetrans*. *Phytopathology*, **50**, 570–1.

ELLENBY, C. (1964) Haemoglobin in the chromatrope of an insect parasitic nematode. *Nature, Lond.*, **202**, 615–6.

ELLENBY, C. and SMITH, L. (1966) Observations on *Panagrellus redivivus* Goodey 1955. *J. Helminth.*, **11**, 323–30.

ELLIOT, A. (1954) Relationship of ageing, food reserves and infectivity of larvae of *Ascaridia gallinae*. *Expt. Parasit.*, **3**, 307–20.

EL-SHERIF, M. A. and MAI, W. F. (1968) Thermotropic response of *Pratylenchus penetrans* and *Ditylenchus dipsaci*. *Nematologica*, **14**, 5–6.

(1969) Thermotactic response of some plant parasitic nematodes. *J. Nematology*, **1**, 43–8.

ERA, E. (1959) Experimental studies on the periodicity of microfilariae. *Endem Dis. Bull. Nagasaki Univ.*, **3**, 252–77.

ESSER, R. P. (1963) Nematode interactions in plates of non-sterile water agar. *Proc. Soil Sci. Fla.*, **23**, 122–38.

ETHERTON, B. and HIGGINBOTHAM, N. (1960) Transmembrane potential measurements of cells of higher plants as related to salt uptake. *Science*, **131**, 409.

EWER, D. W. and BURSELL, E. (1950) A note on the classification of elementary behavioural patterns. *Behaviour*, **3**, 40–7.

FAIRBAIRN, D. (1960) *See* SASSER, J. N. and JENKINS, W. R. (1960).

FAUST, E. C. (1924) The reactions of the miracidium of *Schistosoma japonicum* and *S. haematobium*, in the presence of their intermediate host. *J. Parasit.*, **10**, 199–204.

FRAENKEL, G. S. and GUNN, D. L. (1940) *The Orientation of Animals*. Oxford University Press (revised 1960), Dover, New York.

FRANKS, M. B. and STOLL, N. R. (1945) The isolation of microfilariae from the blood for use as an antigen. *J. Parasit.*, **31**, 158–62.

FULLEBORN, F. (1924) Ueber Taxis (Tropismus) bei *Strongyloides* und Ankylostomenlarven. *Arch. Schiffs U. Tropenhyg.*, **28**, 144–65.

FULLEBORN, F. (1932) Uber die Taxen und des sonstige Verhalten der infektionsfahigen Larven von *Strongyloides* und *Anculostoma. Bakt. Abt.*, 1. *Originale*, **126**, 161–80.

GADD, C. H. and LOOS, C. A. (1941) Host specialisation of *Anguillulina pratensis* (De Man) 1. Attractiveness of roots. *Ann. Appl. Biol.*, **28**, 372–81.

GERHAERT, E. (1965) The head structures of some tylenchs with special attention to the amphidial apertures. *Nematologica*, **11**, 131–6.

GHARIB, H. M. (1955) Observations on the skin penetration by the infective larvae of *Nippostrongylus brasiliensis. J. Helminth.*, **29**, 33–6.

GIBSON, Q. H. and AINSWORTH, S. (1953) Photosensitivity of haem compounds. *Nature, Lond.*, **180**, 1416–17.

GODFREY, G. H. and HOSHINO, H. M. (1933) Studies on certain environmental reactions of the root knot nematode, *Heterodera radicola. Phytopathology*, **23**, 41–62.

GOFMAN-KADOSHNIKOV, P. B., KHOROSCHICHO, E. V. and SMIRNOV, M. I. (1955) The significance of chemical factors on the migration of nematodes. *Dokladi. Akad. Nauk. SSSR.*, **103**, 1127–30 (in Russian).

GOLDSCHMIDT, R. (1903) Histologische Untersuchungen an Nematoden. I. Die Sinnesorgane von *Ascaris lumbricoides* L. and *A. megalocephala. Cloqu. Zool. Jahrb. Abst. Anat.*, **18**, 1–57.

GOODEY, T. (1925) Observations on certain conditions requisite for skin penetration by the infective larvae of *Strongyloides* and *Ancylostoma. J. Helminth.*, **3**, 51–62.

GOODEY, J. B. (1951) The 'hemizonid' a hitherto unreported structure in members of the Tylenchoidea. *J. Helminth.*, **25**, 33–6.

GORDON, R. M. and LUMSDEN, W. H. P. (1939) A study of the behaviour of the mouth parts of the mosquito when taking up blood from living tissues: together with some observations on the ingestion of microfilariae. *Ann. trop. Med. Parasit.*, **23**, 259–78.

GRAY, J. (1953) Undulatory propulsion. *Q. Jl. microsc. Sci.*, **94**, 551–78.

GRAY, J. and LISSMANN, H. W. (1964) The locomotion of nematodes. *J. exp. Biol.*, **41**, 135–54.

GREBECKI, A. (1961) Electrobiological concepts of galvano-taxis in *Paramecium caudata. Progress in Protozoology. Intern. Cong. Prague*, 240–1.

GREEN, C. D. (1966) Orientation of male *Heterodera rostochiensis* Woll. and *Heterodera schachtii* to their females. *Ann. Appl. Biol.*, **58**, 327–39.

GREET, D. N. (1964) Observations on sexual attraction and copulation in the nematode, *Panagrolaimus rigidus*, Schneider. *Nature, Lond.*, **204**, 96.

GUNN, D. L., KENNEDY, J. S. and PIELOU, P. P. (1937) Classification of taxes and kineses. *Nature, Lond.*, **140**, 1064.

GUPTA, S. P. (1962) Galvanotactic reactions of infective larvae of *Trichostrongylus retortaeformis. Expl. Parasit.*, **12**, 118–19.

(1963) Mode of infection and biology of infective larvae of *Molineus barbatus*, Chandler 1942. *Expl. Parasit.*, **13**, 252–5.

HARLEY, G. W. (1932) A theory regarding the role of insect saliva in filarial periodicity. *Trans. R. Soc. trop. Med. Hyg.*, **25**, 487–91.

HARRIS, J. E. and CROFTON, H. D. (1957) Structure and function in the nematodes: internal pressure and cuticular structure in *Ascaris. J. exp. Biol.*, **34**, 116–30.

HAWKING, F. (1953) The periodicity of microfilariae. III. Transfusion of microfilariae into a clean host. *Trans. R. Soc. Trop. Med. Hyg.*, **47**, 82–3.

(1962) Microfilarial infestation as an instance of periodic phenomena seen in a host-parasite relationship. *Ann. N.Y. Acad. Sci.*, **98**, 940–53.

(1965) Advances in filariasis especially concerning periodicity of microfilariae. *Trans. R. Soc. Trop. Med. Hyg.*, **59**, 9–21.

HAWKING, F., SEWELL, P. and THURSTON, J. P. (1950) The mode of action of hetrazan on filarial worms. *Br. J. Pharmac. Chemother.*, **5**, 217–38.

HAWKING, F., WORMS, M. J. and WALKER, P. J. (1965) The periodicity of microfilariae. IX Transfusion of microfilariae (*Edesonfilaria*) into monkeys at a different phase of the circadian rhythm. *Trans. R. Soc. trop. Med. Hyg.*, **59**, 26–41.

HECHT, S. (1919) Dark adaptation. *Mya. J. gen. Physiol.*, **1**, 545–58.

HEMMINGS, C. C. (1966) The mechanism of orientation of Roach, *Rutilus rutilus* L. in an odour gradient. *J. exp. Biol.*, **45**, 465–74.

HERTER, K. (1942) Untersuchungen uber de temperatursinn von Warmbluterparasiten. *Z. Parasitenk.*, **12**, 552–91.

HESLING, J. J. (1966) Preliminary experiments on the control of mycrophagus eelworm in mushroom beds with a note on their swarming. *Pl. Path.*, **15**, 163–6.

HESLING, J. J. and WALLACE, H. R. (1961) Observations on the biology of the Chrysanthemum eelworm, *Aphelenchoides ritzemabosi*, in florist chrysanthemum. I. Spread of eelworm infection. *Ann. Appl. Biol.*, **49**, 195–203.

HESSE, A. J. (1923) On the free-living larval stages of the nematode, *Bunostomum trigonocephalum* (Rud.), a parasite of sheep. *J. Helminth.*, **1**, 21–8.

HINMAN, E. C. (1936) Attempted reversal of filarial periodicity of *Dirofilaria immitis*. *Proc. Soc. Exp. Biol. Med.*, **33**, 524–7.

HOLLIS, J. P. (1960) Mechanism of swarming in *Tylenchorhynchus* sp. *Phytopathology*, **50**, 639.

(1962) Nature of swarming in nematodes. *Nature, Lond.*, **193**, 798–9.

HOPE, W. D. (1964) A taxonomic review of the Genus *Thoracostoma*, Marion 1870 (Nematoda: Leptosomatidae) and a study of the histologic morphology of *Thoracostoma californicum*, Steiner and Albin 1933. Univ. Calif., Davis.

HOPE, W. D. and MURPHY, D. G. (1969) *Rhaptothyreus typicus*, an abyssal marine nematode representing a new family of uncertain taxonomic position. *Proc. Biol. Soc. Wash.*, **82**, 81–92.

HOPPER, B. E. and MEYERS, S. P. (1966) Observations on the bionomics of the marine nematode, *Metoncholaimus* sp. *Nature, Lond.*, **209**, 899–900.

HYMAN, L. (1951) *The Invertebrates. III.* McGraw Hill, New York.

IBRAHIM, I. K. A. (1967) Morphological differences between the cuticle of swarming and non-swarming *Tylenchorhynchus martinii*. *Proc. helminth. Soc. Wash.*, **34**, 18–20.

IBRAHIM, I. K. A. and HOLLIS, J. P. (1967) Nematode orientation mechanisms. 1. A method for determination. *Phytopathology*, **57**, 816.

INGLIS, W. G. (1963) 'Campaniform type' organs in nematodes. *Nature, Lond.*, **197**, 618.

ISHIKAWA, M. (1961) The secretory phenomenon of the nerve cell of *Ascaris*. *Jap. J. Parasit.*, **10**, 1–5. (English Summary).

JAHN, T. L. (1961) The mechanisms of ciliary movement. I. Ciliary reversal and activation by electric currents. The Ludloff phenomenon in terms of core and volume conductors. *J. Protozoology*, **8**, 369–80.

JARMAN, M. (1959) Electrical activity in the muscle cells of *Ascaris lumbricoides*. *Nature, Lond.*, **184**, 1244.

JENNINGS, H. S. (1906) *The Behaviour of the Lower Organisms.* Columbia University Press, New York.

JOHNSON, R. N. and VIGLIERCHIO, D. R. (1961) The accumulation of plant parasitic nematodes around carbon dioxide and oxygen. *Proc. helminth. Soc. Wash.*, **28**, 171–4.

JONES, F. G. W. (1960) Some observations and reflections on host finding by plant nematodes. *Meded. Landb. Hoogesch. Gent.*, **25**, 1009–24.

(1962) Nematology Department. *Rep. Rothamsted Exp. Sta.*, 127–36.

JONES, T. P. (1967) Sex attraction and copulation in *Pelodera teres. Nematologica*, **12**, 518–22.

KAMPHE, L. (1959) Zur Physiologie von *Heterodera* larven unter laboratorium bedingungen al, Festobjekte fur Nematizidprufungen. *Ver. IV Int. Pflz. Kongresses, Hamburg* (1957), **1**, 605–11.

(1960) Die raumliche Verteilung des Primarbefalls von *Heterodera schachtii*, Schmidt in den Wirtswurzeln. *Nematologica*, **5**, 18–26.

(1963) Zucht und Verwendung von Nematoden als Versuchstiere, I. *Panagrellus redivivus* Goodey 1945. *Z. Versuchstierk*, **3**, 11–20.

KATZNELSON, H. and HENDERSON, V. C. (1963) Ammonium as an 'attractant' for soil nematodes. *Nature, Lond.* **198**, 907–8.

KATZNELSON, H., ROUATT, J. W. and PAYNE, T. (1955) The liberation of amino acids and reducing compounds by plant roots. *Plant and Soil*, **7**, 35–48.

KEILIN, D. (1959) The problem of anabiosis or latent life, history and current concept. *Proc. Roy. Soc. B.*, **150**, 149–91.

KENNEDY, D. (1960) Neural photoreception in a lamellibranch mollusc. *J. Gen. Physiol.*, **44**, 277–99.

(1963) Physiology of photoreceptor neurons in the abdominal nerve cord of the crayfish. *J. gen. Physiol.*, **46**, 551–72.

KENNEDY, J. S. (1945) Classification and nomenclature of animal behaviour. *Nature, Lond.*, **155**, 178–89.

KERR, A. and VYTHILINGAM, M. K. (1966) Factors influencing the extraction of nematodes from soil. *Nematologica*, **12**, 511–17.

KHALIL, M. (1922) Thermotropism in ancylostome larvae. *Proc. R. Soc. Med.*, **15**, 6–8.

(1959) Anziehung von Collembolen und Nematoden durch Kohlendioxydquellen. *Mitt. schweiz. ent. Ges.*, **32**, 311–16.

KLINGLER, J. (1961) Anziehungsuersuche mit *Ditylenchus dipsaci* unter Berucksichtigung der Wirkung des Kohlendioxyds, des Redoxpotentials und anderer Kaktoren. *Nematologica*, **6**, 69–84.

(1963) Die orientierung von *Ditylenchus dipsaci* in gemessenen Kunstlichen und biologischen CO_2—Gradienten. *Nematologica*, **9**, 185–99.

(1965) On the orientation of plant nematodes and of some other soil animals. *Nematologica*, **11**, 4–18.

KOEN, H. (1966) The influence of seasonal variations on the vertical distribution of *Meloidogyne javanica* in sandy soils. *Nematologica*, **12**, 297–301.

KUHN, H. (1959) Zum problem der wirtsfindung phytopathogener nematoden. *Nematologica*, **4**, 161–71.

LANE, C. (1930) Behaviour of infective hookworm larvae. *Ann. trop. Med. Parasit.*, **24**, 411–21.

(1933) The taxes of infective hookworm larvae. *Ibid.*, **27**, 237–50.

LEE, D. L. (1965) *The Physiology of Nematodes.* Oliver and Boyd, London.

LEE, D. L. and SMITH, M. H. (1965) Haemoglobin of parasitic animals. *Expl. Parasit.*, **16**, 392–424.

LEES, A. D. (1953) The sensory physiology of the sheep tick, *Ixodes ricinus. J. exp. Biol.*, **25**, 145–207.

LEES, E. (1953) An investigation into the method of dispersal of *Panagrellus silusiae* with particular reference to its desiccation resistance. *J. Helminth.*, **27**, 95–103.

LINFORD, M. B. (1939) Attractiveness of roots and excised shoot tissues to certain nematodes. *Proc. helminth. Soc. Wash.*, **6**, 11–18.

(1942) The transient feeding of root-knot nematode larvae. *Phytopathology*, **32**, 580–9.

LOEWENBERG, J. R., SULLIVAN, T. and SCHUSTER, M. L. (1960) Gall induction by *Meloidogyne incognita incognita* by surface feeding and factors affecting the behaviour pattern of the second stage larvae. *Phytopathology*, **50**, 322–3.

LOOSS, A. (1911) The anatomy and life history of *Ancylostoma duodenale*. Dub. Pt. II *Cairo Records Eg. Govt. Sch. Med.*, **4**, 167–607.

LORENZ, K. Z. (1950) The comparative method in studying innate behaviour patterns. Physiological mechanisms in animal behaviour. *Symp. Soc. Exp. Biol. IV*, Cambridge University Press.

LOWNSBERY, B. F. and VIGLIERCHIO, D. R. (1960) Mechanism of accumulation of *Meloidogyne incognita acrita* around tomato seedlings. *Phytopathology*, **50**, 178–9.

(1961) Importance of response of *Meloidogyne hapla* to an agent from germinating tomato seeds. *Phytopathology*, **51**, 219–22.

LUC, M. (1961) Note préliminaire sur le déplacement de *Hemicycliophora paradoxa* Luc (Nematoda, Criconematidae) dans le sol. *Nematologica*, **6**, 95–106.

LUCKER, J. (1936) Extent of vertical migration of horse strongyle larvae in soils of different type. *J. Agric. Res.*, **52**, 353–61.

(1938) Vertical migrations, distrubution and survival of infective horse strongyle larvae, developing in faeces buried in different soils. *J. Agric. Res.*, **57**, 335–48.

LUND, E. J. and KENYON, W. A. (1927) Relation between continuous bio-electric currents and cell respiration. *J. Exp. Zool.*, **48**, 333–57.

LUNDEGARDH, H. (1940) Investigations as to the absorption and accumulation of inorganic ions. *Lantdr. Hogsk. Ann. Sweden.*, **8**, 234–404.

(1942) Electrochemical relations between the root system and the soil. *Soil Sci.*, **54**, 177–89.

MAGGENTI, A. R. (1964) Morphology of somatic setae: *Thoracostoma californicum* (Nemata: Enoplidae). *Proc. helminth. Soc. Wash.*, **31**, 159–66.

MANSON, P. (1893) *The Filaria sanguinis hominis and certain new forms of parasitic disease in India, China and warm countries.* Lewis, H. K., London.

McBETH, C. W., WHITE, L. V. and ICHIKAWA, S. T. (1964) A new residual nematocide, 2, 4-Dichlorophenyl Methanesulphonate. *Pl. Dis. Reptr.*, **48**, 634–5.

McBRIDE, J. M. (1964) Studies on the induction of swarming in *Tylenchorhynchus martini*, Fielding 1956 (Nematoda: Tylenchida). Ph.D. dissertation, Louisiana State University, Baton Rouge.

McBRIDE, J. M. and HOLLIS, J. P. (1966) The phenomenon of swarming in nematodes. *Nature, Lond.*, **211**, 545–6.

McCUE, J. F. and THORSON, R. E. (1964) Behaviour of parasitic stages of helminths in a thermal gradient. *J. Parasit.*, **50**, 67-71.

(1965) Host effects on the migration of *Nippostrongylus brasiliensis* in a thermal gradient. *J. Parasit.*, **51**, 414–17.

McLAREN, D. J. (1969) Ciliary structures in the microfilaria of *Loa loa. Trans. R. Soc. trop. Med. Hyg.*, **63**, 290–1.

(1970) Preliminary observations on the sensory structures in adult Filariae. *ibid*, **64**, 191–2.

MEYERS, S. P., FEDER, W. A. and BUE, K. M. (1963) Nutritional relationships among certain filamentous fungi and a marine nematode. *Science*, **149**, 520–3.

MEYERS, S. P. and HOPPER, E. (1966) Attraction of the marine nematode, *Metoncholaimus* sp. to fungal substrates. *Bull. Marine Science*, **16**, 142–50.

MONNIG, H. O. (1930) Bionomics of *Trichostrongylus. Union S. Africa Dept. Agric.* 16 *Report Dur. Vet. Res. Anim. Husbandry*, 175–98.

(1941) Measles in cattle and pigs, ways of infection. *J. S. Afr. Vet. Med. Ass.*, **12**, 59–61.

MORETON, B. D. (1956) Problems of mushroom eelworms. *The Grower*, Aug. 25.

MORGAN, D. O. (1928) On the infective larvae of *Ostertagia circumcincta* Stadelmann (1894), a stomach worm of sheep. *J. Helminth.*, **6**, 183–92.

MORTON, J. E. (1962) Habit and orientation in the small commensal bivalve mollusc, *Montacuta ferruginosa*. *Anim. Behav.*, **10**, 126–33.

MUELLER, T. F. (1961) The laboratory propagation of *Spirometra mansonoides* as an experimental tool. V. Behaviour of the sparganum in and out of the mouse host and the formation of the immune precipitates. *J. Parasit.*, **47**, 879–83.

MURPHY, D. G. (1963) A note on the structure of the nematode ocelli. *Proc. helminth. Soc. Wash.*, **30**, 25.

NEKIPELOVA, R. A. (1956) Problem parasitologii. *Trans. Scient. Cong. Parasit. Ukrainian S.S.R.*, **2**.

NELSON, G. S. (1964) Factors influencing the development and behaviour of filarial nematodes in their arthropodan host, in Host Parasite relationships of invertebrate hosts. Edit. A. E. R. Taylor, Blackwell, Oxford.

NEWELL, R. C. (1966) Effect of temperature on the metabolism of Poikilotherms. *Nature, Lond.*, **212**, 426–28.

NIELSON, C. O. (1961) Respiratory metabolism of some populations of enchytraeid worms and free-living nematodes. *Oikos*, **12**, 17–35.

(1967) Nematodes. *Soil Biology*. Ed. Burges, A. and Raw, F., Academic Press, London.

NIKOLSKY, G. V. (1963) *The Ecology of Fishes*. Academic Press, London.

O'CONNER, F. W. and BEATTY, H. A. (1937) The abstraction by *Culex fatigans* of *Microfilaria bancrofti* from man. *J. trop. Med. Hyg.*, **40**, 101–3.

PANTIN, C. F. A. (1950) Behaviour patterns of Lower Invertebrates. *Symp. Soc. Exp. Biol. IV*. Cambridge University Press.

PARAMONOV, A. A. (1954) On the structure and function of phasmids. *Trud. Gelm. Lab. Akad. Nauk. S.S.S.R.*, **7**, 19–49. (in Russian).

PARKER, J. C. and HALEY, A. J. (1960) Phototactic and thermotactic responses of the filariform larvae of the rat nematode, *Nippostrongylus muris*. *Expl. Parasit.*, **9**, 92–7.

PASSANO, L. M. (1963) The origin of the nervous system. *Proc. XVI Intern. Cong. Zool.*, **2**, Washington D.C.

(1963a) Pacemaker hierarchies controlling the behaviour of Hydras. *Nature, Lond.*, **199**, 1174–5.

PAYNE, F. K. (1922) Investigations on the control of hookworm diseases XI. Vertical migration of infective hookworm larvae in the soil. *Amer. J. Hyg.*, **2**, 254–63.

(1923) Investigations on the control of hookworm disease. XIV. Field experiments on vertical migration of hookworm larvae. *Ibid.*, **3**, 547–83.

(1923a) Investigations on the control of hookworm disease. XXX. Studies on factors involved in migration of hookworm larvae in soil. *Ibid.*, **3**, 547–87.

PEACOCK, F. C. (1959) The development of a technique for studying the host–parasite relationship of the root knot nematode, *Meloidogyne incognita*, under controlled conditions. *Nematologica*, **4**, 43–55.

(1961) A note on the attractiveness of roots to plant parasitic nematodes. *Nematologica*, **6**, 85–6.

PETERS, B. G. (1928) On the bionomics of the vinegar eelworm. *J. Helminth.*, **6**, 1–38.

(1952) Toxicity tests with vinegar eelworm. I. Counting and culturing. *J. Helminth.*, **26**, 97–110.

PICTON, H. D. (1966) Some responses of *Drosophila* to weak magnetic and electrostatic fields. *Nature, Lond.*, **211**, 303–4.

PITCHER, R. S. (1967) The host–parasite relations, and ecology of *Trichodorus viruliferus* on apple roots, as observed from an underground laboratory. *Nematologica*, **13**, 547–57.

PROSSER, C. L., BISHOP, D. W., BROWN, F. A., JAHN, R. L. and WULFF, V. J. (1950) *Comparative Animal Physiology*, Saunders, London.

RAKSHPAL, R. (1959) On the behaviour of the pigeon louse, *Columbicola columbae* L. (Mallophaga). *Parasitology*, **49**, 232–41.

RAUTHER, M. (1907) Uber den Ban des Oesophagus und lie lokalisation der Nierenfunktion bei freilebenden Nematoden. *Zool. Jahrb. Anat. Bd.*, **23**.

REED, E. M. and WALLACE, H. R. (1965) Leaping locomotion by an insect parasitic nematode. *Nature, Lond.*, **206**, 210–11.

REES, G. (1950) Observations on the vertical migration of the L3 of *Haemonchus contortus* on experimental plots of *Lolium perenne* in relation to meteorological and micrometeorological factors. *Parasitology*, **40**, 127–42.

REESAL, M. R. (1951) Observations on the biology of the infective larvae of *strongyloides agouti*. *Can. J. Zool.*, **29**, 109–15.

REMANE, A. (1933) Verteilung und Organisation der Genthonischen Mikrofauna der Kieler Bucht. *Wiss. Meeresunters Kiel N.F.*, **21**.

RENSCH, B. (1925) Zwie quantitative reizphysiologische untersuchgensmethoden fur Rubernematoden. *Ztschr. Wiss. Zool.*, **123**.

ROBERTS, L. S. AND FAIRBAIRN, D. (1965) Metabolic studies on adult *Nippostrongylus brasiliensis* (Nematoda: Trichostrongyloidea). *J. Parasit.*, **51**, 129–38.

ROBINSON, J. (1962) *Pilobolus* spp. and the translocation of the infective larvae of *Dictyocaulus viviparus* from faeces to pasture. *Nature, Lond.*, **193**, 353–4.

ROBINSON, J., POYNTER, D. and TERRY, R. J. (1962) The role of the fungus *Pilobolus* in the spread of the infective larvae of *Dictyocaulus viviparus*. *Parasitology*, **52**, 17–18.

ROCHE, M. (1966) Influence of male and female *Ancylostoma caninum* on each other's distribution in the intestine of the dog. *Expl. Parasit.*, **9**, 250–6.

ROCHE, M. and TORRES, C. M. (1960) A method for the *in vitro* study of hookworm activity. *Exp. Parasit.*, **19**, 327–31.

RODE, H. (1962) Untersuchungen uber des Wander ver mogen von Larven des Kartoffel nematoden (*Heterodera rostochiensis* Woll.) in Modelluersuchen mit verschiedenen Bodenarten. *Nematologica*, **7**, 74–82.

RODE, H. (1965) Uber einige Methoden zur Verhaltensforschung bei Nematoden. *Pedobiologia*, **5**, 1–16.

RODE, H. and STARR, G. (1961) Die photographische darstellung der kriechspuren (Ichnogramme) von Nematoden und ihre Loedeutung. *Nematologica*, **6**, 266–71.

ROGERS, W. P. (1939) The physiological ageing of ancylostome larvae. *J. Helminth.*, **17**, 195–202.

(1940) The physiological ageing of the infective larvae of *Haemonchus contortus*. *J. Helminth.*, **18**, 183–92.

(1940a) The effect of environmental conditions on the accessibility of third stage trichostrongyle larvae to grazing animals. *Parasitology*, **32**, 208–26.

(1948) The respiratory metabolism of parasitic nematodes. *Parasitology*, **39**, 105–9.

(1949) The biological significance of haemoglobin in nematode parasites. I. The characteristics of the purified pigments. *Aust. J. Sci. Res. B.*, **2**, 287–303.

(1960) The physiology of the infective process of nematode parasites: the stimulus from the host. *Proc. R. Soc.*, *B*, **152**, 367–86.

(1962) *The Nature of Parasitism*. Academic Press.

(1966) Reversible inhibition of a receptor governing infection with some nematodes. *Expl. Parasit.*, **19**, 15–20.

(1966a) The reversible inhibition of exsheathment in some parasitic nematodes. *Comp. Biochem. Physiol.*, **17**, 11–3–1110.

ROGERS, W. P. and SOMMERVILLE, R. I. (1963) The infective stage of nematode parasites, and its significance in parasitism. *Advances in Parasitology*, **1**, 109–78. Academic Press.

ROGGEN, D. R., RASKI, D. J. and JONES, N. O. (1966) Cilia in nematode sensory organs. *Science*, **152**, 515–16.

ROHDE, R. A. (1960) The influence of carbon dioxide on respiration of certain plant-parasitic nematodes. *Proc. helminth. Soc. Wash.*, **27**, 160–4.

RONALD, K. (1960) The effects of physical stimuli on the larvae of *Terranova decipiens*. I. Temperature. *Can. J. Zool.*, **38**, 623–42.

(1962) The effects of physical stimuli on the larvae of *Terranova decipiens*. II. Relative humidity, gases and pressure. *Ibid.*, **40**, 1223–7.

(1963) The effects of physical stimuli on the larvae of *Terranova dεcipiens*. III. Electromagnetic spectrum and galvanotaxes. *Ibid.*, **41**, 197–217.

ROSE, J. H. (1957) Observations on the bionomics of the free-living first stage larvae of the sheep lungworm, *Muellerius capillaris*. *J. Helminth.*, **31**, 17–28.

(1963) Observations on the free-living stages of the stomach worm *Haemonchus contortus*. *Parasitology*, **53**, 469–82.

ROSS, M. M. R. (1967) Modified cilia in sensory organs of juvenile stages of a parasitic nematode. *Science*, **156**, 1494–5.

ROTHSCHILD, M. (1962) Changes in behaviour of the intermediate hosts of trematodes. *Nature, Lond.*, **193**, 1312–13.

SAITO, F. (1960) Studies on the swimming speed of cercariae of *Schistosoma japonicum*. *J. Kumamoto Med. Soc.*, **34**, 2158–79.

SANDSTEDT, R. and SCHUSTER, M. L. (1962) Liquid trapping of *Meloidogyne incognita incognita* about roots in agar medium. *Phytopathology*, **52**, 174–5.

SANDSTEDT, R., SULLIVAN, T. and SCHUSTER, M. L. (1961) Nematode tracks in the study of movement of *Meloidogyne incognita incognita*. *Nematologica*, **6**, 261–5.

SANTMEYER, P. H. (1956) Studies on the metabolism of *Panagrellus redivivus* (Nematoda, Cephalobidae). *Proc. helminth. Soc. Wash.*, **23**, 30–6.

SASSER, J. N. and JENKINS, W. R. (1960) *Nematology, fundamentals and recent advances with emphasis on plant parasitic and soil forms*. University North Carolina, Chapel Press.

SAWADA, J. (1961) Biological studies on the third stage larvae of canine hookworm. *Jap. J. Parasit.*, **10**, 398–409.

SCHULTZ, E. (1931) Die Augen Freilebender Nematoden. *Zool. Anz.*, 95–6.

SCHUSTER, M. L. and SANDSTEDT, R. M. (1962) Studies on host attraction and symptom induction by *Meloidogyne* in tissue culture. *Nematologica*, **7**, 8–10.

SELIGER, A. H. and MCELROY, W. P. (1965) *Light: Physical and Biological Action*. Academic Press, London.

SHEPHERD, A. M. (1959) The invasion and development of some species of *Heterodera* in plants of different host status. *Nematologica*, **4**, 253–67.

(1961) Hatching and hatching factors. In Jones, F. G. W. *Nematology Dept. Rep. Rothamsted. exp. Stat.*, 127–8.

SLACK, D. A. and HAMBLEN, M. L. (1961) The effects of various factors on larval emergence from cysts of *Heterodera glycines*. *Phytopathology*, **51**, 350–5.

SMYTH, J. D. (1966) *The Physiology of Trematodes*. Oliver and Boyd, London.

SOLIMAN, K. N. (1953) Studies on the bionomics of the preparasitic stages of *Dictyocaulus viviparus* with a reference to the same in the allied species of sheep, *D. filaria*. *Br. vet. J.*, **109**, 364–81.

SOMMERVILLE, R. I. (1964) The effect of carbon dioxide on the development of third stage larvae of *Haemonchus contortus in vitro*. *Nature, Lond.*, **202**, 316–17.

SOULAIRAC, A. (1949) Classification des reactions d'orientation des animaux (tropisms). *Année Biol.*, **25**, 1–14.

SPINDLER, L. (1934) Field and laboratory studies on the behaviour of the larvae of the swine kidney worm. *U.S. Dept. Agric. Tech. Bull.*, **405**.

SPRENT, J. F. A. (1946) Some observations on the bionomics of *Bunostomum phlebotomum*, a hookworm of cattle. *Parasitology*, **37**, 202–10.

STAAR, G. (1959) Veber sinige Ergebnisse aus reizphysioloschen Versuchen an *Heterodera rostochiensis*. *Deutsche Akad. Landıvortschaftswias*, Berlin, tangungster Nr. 20.

STANILAND, L. N. (1957) The swarming of Rhabditid eelworms in mushroom houses. *Pl. Path.*, **6**, 61–2.

(1959) *Plant Nematology*, Tech. Bull. 7, Min. of Agric. & Fish., H.M. Stationery Office.

STAUFFER, H. (1925) Die Lokomotion der Nematoden. *Zool. Jb.* (abt. 1), **49**, 1–118.

STEINER, G. (1925) The problem of host selection and host specialization of certain plant-infesting nemas and its application in the study of nemic pests. *Phytopathology*, **15**, 499–534.

(1955) Structure, function and host in root-knot nematodes. *Phytopathology*, **45**, 466.

STEINER, G. and ALBIN, F. T. (1946) Resuscitation of the nematode, *Tylenchus polyhypnus* n.sp. after almost 39 years dormancy. *J. Wash. Acad. Sci.*, **36**, 97–9.

STEVEN, D. M. (1963) The dermal light sense. *Biol. Rev.*, **38**, 204–40.

STEWART, M. A. and DOUGLAS, J. R. (1938) Studies on the bionomics of *Trichostrongylus axei* (Cobbold) and its seasonal incidence on irrigated pastures. *Parasitology*, **30**, 477–90.

STREET, M. W., JR. and BRANCH, SR. K. W. (1964) The reaction of *Australorbis glabratus* to pulsating direct current. *J. Parasit.*, **50**, 589–90.

STREU, H. T., JENKINS, W. R. and HUTCHINSON, M. T. (1961) Nematodes associated with carnations. *New Jersey Agric. Exp. Sta. Rutgers Bull.*, **800**.

STRONG, R. P., SANDGROUND, J. H., BEQUAERT, J. C. and OCHOA, M. M. (1934) *Onchocerciasis*, Harvard University Press, Cambridge, Mass.

TAKAHASHI, T., MORI, K. and SHIGETA, Y. (1961) Phototactic, thermotactic and geotactic responses of miracidia of *Schistosoma japonicum. Jap. J. Parasit.*, **10**, 686–91.

TANIGUCHI, R. (1933) Notes on the chemotactic response of *Rhabditis filiformis*, Butschlii. *Proc. Imperial Acad.*, **9**, 432–5.

TAYLOR, G. I. (1951) Analysis of the swimming of microscopic organisms. *Proc. R. Soc., A*, **209**, 447–61.

(1952) Analysis of swimming of long and narrow animals. *Proc. R. Soc., A*, **214**, 158–83.

THOMAS, H. A. (1959) On *Criconemoides xenoplax*, Raski, with special reference to its biology under laboratory conditions. *Proc. helminth. Soc. Wash.*, **26**, 55–9.

THOMASON, J., VAN GUNDY, S. D. and KIRKPATRICK, J. D. (1964) Motility and infectivity of *Meloidogyne javanica* as affected by storage time and temperature in water. *Phytopathology*, **54**, 192–5.

THORPE, H. W., CROMBIE, A. C., BILL, R. and DARRAH, J. H. (1947) The behaviour of wireworms in response to chemical stimulation. *J. Exp. Biol.*, **23**, 234–66.

THORSON, R. E., MUELLER, J. F. and MCCUE, J. F. (1964) Thermotactic responses of *Spirometra* plerocercoids. *J. Parasit.*, **50**, 529–30.

TIMM, R. W. (1951) A new species of marine nema, *Thoracostoma magnificum* with a note on the possible 'pigment cell' nuclei of the ocelli. *J. Wash. Acad. Sci.*, **41**, 331–38.

(1952) A survey of the marine nematodes of Chesapeake Bay, Maryland. *State Maryland Board of Natural Resources, Dept. of Research and Education. Publ.* **95**.

TINBERGEN, N. (1942) An objectivistic study of the innate behaviour of animals. *Bibliotheca biotheoretical., V*, 39–98.

VAN DURME, J. (1902) Quelques notes sur les embryons de *Strongyloides intestinalis* et leur penetration par le peau. *Thompson Yates Lab. Rep.*, **4**, 741.

VAN GUNDY, S. D. (1961) *The sheath nematode* (film). Dept. Nematology, University California, Riverside.

(1965) Nematode behaviour. *Nematologica*, **11**, 19–32.

(1965a) Factors in survival of nematodes. *Ann. Rev. Phytopathol.*, **3**, 43–68.

VAN GUNDY, S. D., BIRD, A. F. and WALLACE, H. R. (1967) Ageing and starvation in larvae of *Meloidogyne javanica* and *Tylenchulus semipenetrans*. *Phytopath.*, **57**, 559–71.

VEKI, T. (1957) Experimental studies on the third stage larvae of *Gnathostoma spinigerum*. *Igaku Kenkyu Fukuoka*, **27**, 1162–96.

VIAUD, G. (1940) Researches experimentals le phototropism de Rotiferes. *Bull. Biol.*, **74**, 249–308.

VIGLIERCHIO, D. R. (1961) Attraction of parasitic nematodes by plant root emanations. *Phytopathology*, **51**, 136–42.

VIGLIERCHIO, D. R. and CROLL, N. A. (1969) The comparative effects of chloramines on a range of nematodes. *J. Nematology*, **1**, 35–9.

VON BRAND, T. (1960) Influence of size, motility, starvation, and age on metabolic rate. In Sasser, J. and Jenkins, W. R., 233–41.

VOSS, W. (1930) Beitrage zur Kenntnis der Alchenkankhait der Chrysanthemum. *Stschr. Parasitenk.*, **2**, 310–56.

WAKESHIMA, T. (1933) Experimental studies on the tropisms of mature larvae in hookworms. IV. Report phototropism, thermotropism, and barytropism of the mature larvae of *Ancylostoma caninum*. *Taiwan Igakkai Zashi*, **32**, 152–3. (English summary.)

WALD, G. (1946) Chemical evolution of vision. *Harvey Soc. Lecture*, **41**, 117–60.

WALLACE, H. R. (1955) Factors influencing the emergence of larvae from cysts of the beet eelworm. *Heterodera schachtii*. *J. Helminth.*, **29**, 3–16.

(1956) Migration of nematodes. *Nature, Lond.*, **177**, 287–8.

(1958) Movement of eelworms. I. The influence of pore size and moisture content of the soil on the migration of larvae of the beet eelworm, *Heterodera schachtii*, Schmidt. *Ann. appl. Biol.*, **46**, 74–85.

(1958a) Movement of eelworms. II. A comparative study of movement soil of *Heterodera* Schmidt and of *Ditylenchus dipsaci* (Kuhn) Filipjev. *Ann. appl. Biol.*, **46**, 86–94.

(1958b) Movement of eelworms. III. The relationship between length, activity and mobility. *Ann. appl. Biol.*, **46**, 662–8.

(1958c) Observations on the emergence from cysts and the orientation of larvae of three species of the genus *Heterodera* in the presence of host plant roots. *Nematologica*, **3**, 236–43.

(1959) Movement of eelworms. IV. The influence of water percolation. *Ann. appl. Biol.*, **47**, 131–9.

(1959a) Movement of eelworms. V. Observations on *Aphelenchoides ritzema-bosi* (Schwartz, 1912) Steiner 1932 on florist's chrysanthemums. *Ann. appl. Biol.*, **47**, 350–60.

(1959b) The movement of eelworms in water films. *Ann. appl. Biol.*, **47**, 366–70.

(1960) Movement of eelworms. VI. The influence of soil type, moisture gradients and host plant roots on the migration of the potato-root eelworm *Heterodera rostochiensis*, Wollenweber. *Ann. appl. Biol.*, **48**, 107–20.

(1960a) Observations on the behaviour of *Aphelenchoides ritzema-bosi* in chrysanthemum leaves. *Nematologica*, **5**, 315–21.

(1961) The orientation of *Ditylenchus dipsaci* to physical stimuli. *Nematologica*, **6**, 222–36.

(1961a) The bionomics of the free-living stages of zoo-parasitic and phyto-parasitic nematodes—a critical survey. *Helminth. Abs.*, **30**, 1-22.

(1961b) Factors influencing the ability of *Heterodera* larvae to reach host plant roots. *Recent. Advances in Botany*, **5**, 407–10.

(1962) Observations on the behaviour of *Ditylenchus dipsaci* in soil. *Nematologica*, **7**, 91–101.

(1963) *The Biology of Plant Parasitic Nematodes*. Edward Arnold, London.

(1966) Factors influencing the infectivity of plant parasitic nematodes. *Proc. Roy. Soc., B*, **164**, 592–614.

(1968) Undulatory locomotion of the plant parasitic nematode *Meloidogyne javanica. Parasitology*, **58**, 377–91.

(1968a) The Dynamics of Nematode Movement. *Ann. Rev. Phytopathology*, 6., 91–114.

WALLACE, H. R. and DONCASTER, C. C. (1964) A comparative study of the movement of some microphagous, plant parasitic and animal parasitic nematodes. *Parasitology*, **54**, 313–26.

WANG, C. F., LIN, C. L. and CH'EN, W. H. (1959) The mechanism of microfilarial periodicity. *Chinese Medical Journal*, Peking, **78**, 171–2.

WEBSTER, J. M. (1964) The effect of storage conditions on the infectivity of narcissus stem eelworms. *Nature, Lond.*, **202**, 571–5.

WEISCHER, B. (1959) Experimentelle untersuchungen uber die Wanderung von Nematoden. *Nematologica*, **4**, 172–86.

WEISER, W. (1955) The attractiveness of plants to larvae of root-knot nematodes. I. The effect of tomato seedlings and excised roots on *Meloidogyne hapla* Chitwood. *Proc. helminth. Soc. Wash.*, **22**, 106–12.

(1956) The attractiveness of plants to larvae of root-knot nematode. II. The effect of excised bean, eggp-lant and soybean on *Meloidogyne hapla* Chitwood. *Proc. helminth. Soc. Wash.*, **23**, 59–64.

(1959) Free-living marine nematodes. IV. General Part. Report of Lund University Chile Expedit., 1948–9. *K. fysiogr. Sallsk. Lund Handl., n.f.* 70.

WHITEHEAD, A. G. and HEMMING, J. R. (1965) A comparison of some quantitative methods of extracting small vermiform nematodes from soil. *Ann. appl. Biol.*, **55**, 25–38.

WHITTAKER, F. W. (1969) Galvanotaxis of *Pelodera strongyloides* (Nematoda, Rhabditidae). *Proc. Helminth. Soc. Wash.*, **36**, 40.

WIDDOWSON, E., DONCASTER, C. C. and FENWICK, D. W. (1958) Observations on the development of *Heterodera rostochiensis* Woll., in sterile root cultures. *Nematologica*, **3**, 308–14.

WIGGLESWORTH, V. B. and GILLETT, I. D. (1934) The functions of the antennae in *Rhodnius prolixus* (Hemiptera) and the mechanism of orientation to the host. *J. exp. Biol.*, **11**, 120–39.

WILSON, P. A. G. (1957) Studies on the physiology and behaviour of the immature stages of *Trichstrongylus retortaeformis* (Zeder 1800). Ph.D. Thesis, University of Leeds.

(1965) The effect of temperature change on the oxygen uptake of the infective larvae of *Nippostrongylus braziliensis. Expl. Parasit.*, **17**, 318–25.

(1966) The light sense in nematodes. *Science*, **151**, 337–8.

WRIGHT, C. A. (1959) Host location by trematode miracidia. *Ann. trop. Med. Parasit.*, **53**, 288–92.

WRIGHT, K. A. and HOPE, W. D. (1968) Elaborations of the cuticle of *Acanthonchus duplicatus* Weiser 1959, Nematoda: Cyatholaimidae as revealed by light and electron microscopy *Canad. J. Zool.*, **46** (5), 1005–11.

YASURACKA, K. (1953) Ecology of miracidia. I. On the perpendicular distribution and rheotaxis of the miracidia of *Fasciola hepatica* in water. *Jap. J. Med. Sci. Biol.*, **6**, 1–10.

(1954) Ecology of miracidia. II. On the behaviour to light of the miracidium of *Fasciola hepatica* in water. *ibid.*, **7**, 181–92.

YOELI, M. (1957) Observations on the agglutination and thigmotaxis of microfilariae in bancroftian filariasis. *Trans. R. Soc. trop. Med.*, **51**, 132–6.

YOSHIDA, M. and MILLOTT, N. (1960) The shadow reaction of *Diadema antillarum*. 3. Re-examination of the spectral sensitivity. *J. exp. Biol.*, **12**, 229–38.

YOUNG, J. Z. (1935) The photoreceptors of lampreys. *J. exp. Biol.*, **12**, 229–38.

NEMATODE INDEX

(AP animal parasitic; PP plant parasitic; M marine; FL freeliving in soil or fresh water.)

SUBJECT INDEX